## Praise for *Herbal Supplements and the Brain*

"Both skeptics and believers in the value of herbal supplements for brain conditions will enjoy the calm objective analysis to which these two experienced pharmacologists put the most popular products. You may not like their conclusions, but their evidence is convincing."

—**Floyd E. Bloom**, MD, Professor Emeritus, Molecular and Integrative Neuroscience Department, TSRI

"Written with authority yet as lucid and enticing as a novel, the Enna/Norton book is certainly the finest volume I know addressing the interface of herbs, the brain, and behavior. It will be of value and fun for the educated layperson as well as the professional."

—**Solomon H. Snyder**, M.D., Department of Neuroscience, Johns Hopkins University School of Medicine

"It all began with Adam's apple. Knowing what you add to your diet may change your life. Getting a kick from a cup of coffee, fighting depression with St. John's wort, drifting away with valerian, or reaching a ripe old age with Gingko, this book gives insights into the pros and cons of taking herbal supplements. Excellent and entertaining reading!"

—**Hanns Möhler**, Professor of Pharmacology, University of Zurich, Switzerland

"If you are someone who takes and believes in herbal supplements, then this book is a must-read for you. I'll bet you will be surprised at some of the information. It is written by two extraordinarily qualified authors, who have decades of experience with the effects and toxicities of drugs and supplements. The aim of the book is to use proven criteria to evaluate if herbal supplements are effective or not. This information is not always easy to find, so read on."

—**Michael J Kuhar, Ph.D.**, Yerkes National Primate Research Center of Emory University, Candler Professor of Neuropharmacology, School of Medicine, Georgia Research Alliance Eminent Scholar, Center for Ethics of Emory University

# Herbal Supplements and the Brain

# Herbal Supplements and the Brain

## Understanding Their Health Benefits and Hazards

S. J. Enna
Stata Norton

Illustrated by Kevin S. Smith

Vice President, Publisher: Tim Moore
Associate Publisher and Director of Marketing: Amy Neidlinger
Editorial Assistant: Pamela Boland
Development Editor: Russ Hall
Operations Specialist: Jodi Kemper
Assistant Marketing Manager: Megan Graue
Cover Designer: Chuti Prasertsith
Managing Editor: Kristy Hart
Project Editor: Anne Goebel
Copy Editor: Geneil Breeze
Proofreader: Debbie Williams
Indexer: Erika Millen
Compositor: Nonie Ratcliff
Manufacturing Buyer: Dan Uhrig


Publishing as FT Press
Upper Saddle River, New Jersey 07458

FT Press offers excellent discounts on this book when ordered in quantity for bulk purchases or special sales. For more information, please contact U.S. Corporate and Government Sales, 1-800-382-3419, corpsales@pearsontechgroup.com. For sales outside the U.S., please contact International Sales at international@pearson.com.



Printed in the United States of America

First Printing May 2012

ISBN-10: 0-13-282497-3
ISBN-13: 978-0-13-282497-2

Pearson Education LTD.
Pearson Education Australia PTY, Limited.
Pearson Education Singapore, Pte. Ltd.
Pearson Education Asia, Ltd.
Pearson Education Canada, Ltd.
Pearson Educación de Mexico, S.A. de C.V.
Pearson Education—Japan
Pearson Education Malaysia, Pte. Ltd.

*Library of Congress Cataloging-in-Publication Data*

Enna, S. J.
Herbal supplements and the brain : understanding their health benefits and hazards / S.J. Enna, Stata Norton.
p. cm.
Includes bibliographical references and index.
ISBN 978-0-13-282497-2 (hardback : alk. paper) 1. Herbs—Therapeutic use. 2. Alternative medicine. 3. Dietary supplements—Therapeutic use. I. Norton, Stata, 1922- II. Title.
RM666.H33E55 2013
615.3'21—dc23

2012006286

*We thank our spouses, Colleen Enna and David Ringle,*
*for the decades they have devoted to encouraging us*
*to pursue our scientific interests and career goals.*
*Neither this book, nor any of our other accomplishments,*
*would have been possible without their patience, support,*
*and understanding. This work is dedicated to them.*

# Contents

# Acknowledgments

We thank Mr. Kirk Jensen, Mr. Russ Hall, and their colleagues at Pearson for their guidance and assistance in creating this volume. Thanks, too, to Mr. Kevin S. Smith for preparing the illustrations and for contributing his creative talents to this project. We are particularly grateful to Ms. Lynn LeCount at the University of Kansas Medical School for her editorial and technical assistance. Without her efforts, this work would not have been possible.

# About the Authors

A Kansas City native, **S. J. Enna** received his undergraduate education at Rockhurst University and his M.S. and Ph.D. degrees in pharmacology from the University of Missouri. Following postdoctoral work at the University of Texas Southwestern Medical School in Dallas, F. Hoffmann-LaRoche in Basel, Switzerland, and Johns Hopkins University Medical School, Dr. Enna joined the faculty at the University of Texas Medical School in Houston in the departments of Pharmacology and of Neurobiology and Anatomy. In 1986, Dr. Enna was appointed Scientific Director of Nova Pharmaceutical Corporation in Baltimore. At this time, he also held appointments as a Lecturer in Neuroscience at Johns Hopkins Medical School and Adjunct Professor of Pharmacology at Tulane Medical School in New Orleans. Dr. Enna returned to Kansas City in 1992 as Professor and Chair of Pharmacology at the University of Kansas Medical Center, where he is currently Associate Dean for Research and Graduate Education, and Professor of Pharmacology and of Physiology. Dr. Enna has published nearly 300 research reports, reviews, and book chapters describing his research, and has edited or coedited 30 books on various subjects relating to neuropharmacology. For decades, he has served as a scientific consultant for most of the major pharmaceutical companies and as a board member on government panels and private research foundations. He has been appointed to the editorial boards of numerous pharmacology journals and has served as editor of leading journals in the field, including the *Journal of Pharmacology and Experimental Therapeutics.*

At present, Dr. Enna is editor of *Biochemical Pharmacology* and *Pharmacology and Therapeutics*, coeditor of *Current Protocols in Pharmacology*, and series editor of *Advances in Pharmacology.* Dr. Enna has held many elective offices in professional societies, including the presidency of the American Society for Pharmacology and Experimental Therapeutics. Currently, he is Secretary-General of the International Union of Basic and Clinical Pharmacology. He has received numerous awards in recognition of his contributions to the discipline of pharmacology. Among these are two Research Career

Development Awards from the National Institutes of Health, the John Jacob Abel and Torald Sollmann Awards from the American Society for Pharmacology and Experimental Therapeutics, the Daniel H. Efron Award from the American College of Neuorpsychopharmacology, and the PhRMA Foundation Excellence Award from the Pharmaceutical Manufacturer's Association Foundation. Dr. Enna is internationally recognized for his research in neuropharmacology, especially his contributions in characterizing the biochemical and pharmacological properties of neurotransmitter receptors in general, and the γ-aminobutyric acid (GABA) system in particular.

**Dr. Stata Norton** received her B.S. degree in Biology from the University of Connecticut, Storrs, her Master's degree in Zoology from Columbia University in New York, and her Ph.D. degree in Zoology and Physiological Chemistry from the University of Wisconsin in Madison. After working for over a decade as a neuropharmacologist at the Wellcome Research Laboratories in Tuckahoe, New York, in 1962, she accepted a faculty appointment in Department of Pharmacology, Toxicology, and Therapeutics at the University of Kansas Medical School in Kansas City. While at the University of Kansas, she also served as Professor of Dietetics and Nutrition, Dean of the School of Allied Health, and Professor at the Institute for Cell Biology in Valencia, Spain. Since 1990, she has been Emeritus Professor of Pharmacology at the University of Kansas Medical School.

Dr. Norton was the recipient of a Research Scientist Award from the National Institute of Mental Health and was elected to Outstanding Educators of America in 1975. She served on the Editorial Advisory boards for *Psychopharmacology*, *Neurotoxicology*, and *Toxicology and Industrial Health*. She was President of the Central States Chapter of the Society of Toxicology and a member of the Preclinical Psychopharmacology Study Section of the National Institute of Mental Health. She has published more than 150 scientific research articles and reviews in scientific journals and books. Her main research interests are the development of quantitative methods for analyzing animal behavior and the characterization of the effects of chemicals on behavior. For more than two decades, she authored a major review on the toxic effects of plants for the reference work *Toxicology: The Basic Science of Poisons*.

# Preface

While for more than 200,000 years humans have been consuming plant materials, such as flowers, fruits, leaves, and roots, for therapeutic benefit, it is only in the last 150 years that scientists have been able to isolate, identify, examine, and categorize the biologically active constituents in plants. Many of the compounds identified, or chemical derivatives of them, are now employed as drugs. The ability to obtain such precise scientific information, and to synthesize other active compounds, opened the way for legislators in the early twentieth century to enact laws regulating the marketing and sale of chemicals for therapeutic purposes. The creation of these regulatory requirements was spurred by the fact that many inert, and sometimes toxic, products were sold as medications to the public. Current laws mandate that any substance marketed as a treatment for a particular condition must first undergo rigorous testing in laboratory animals and humans to demonstrate its safety and effectiveness.

Although prescription and over-the-counter medications are subject to tight federal oversight, there are few regulations concerning the sale of herbal supplements. In the United States, the chief requirement is that no formal claims be made of any therapeutic benefit resulting from the use of these products. Nonetheless, consumers are continuously exposed in the lay press and online to reports on the purported curative properties of certain herbs or how their consumption can help prevent disease. Such reports are no doubt responsible for driving sales in this multibillion dollar industry. However, like our ancient ancestors, today's consumer may be purchasing and consuming these products for health benefits based solely on the word of others, not as a result of an independent and objective analysis of the data supporting the claims. This is understandable, as most lack the technical background for making an informed scientific judgment. The aim of this book is to address this need.

Herbal products are used around the world for a variety of purposes. Among these is the treatment of central nervous system disorders, such as anxiety, insomnia, alcoholism, dementia, and depression. Herbal supplements are also taken to modify brain function in the treatment of other conditions, such as chronic pain and obesity. Because some of the symptoms of these disorders resolve over time without medication, and many have a strong psychological component, it is often difficult to prove the efficacy of an herbal product as a treatment for these conditions. That is, while the effectiveness of a dietary supplement that reduces body weight would be apparent, the contribution of an herbal product in lessening feelings of depression, or in enhancing cognitive abilities, is more difficult to quantify. For this reason, the claims for such benefits may not be supported by experimental data. In this regard, the consumer may be no different than the primitive who ingested a plant material to alter his mental status. Sometimes it worked; often it did not. A change in perception or feelings, or in sleep patterns, could be the result of an active constituent in the plant, or the power of suggestion. Prolonged consumption of any product with no inherent value is not only costly, but potentially dangerous as anything taken into the body can have toxic consequences. This volume is devoted to a discussion of herbal supplements taken to affect brain function because of the unique challenges associated with assessing the effectiveness of such products.

Written for the nonscientist, the book is informally divided into two parts. The first section, chapters 1–4, provides an historical perspective on the use of plant products to modify central nervous system function and on the development of the techniques employed for drug discovery. Included is a discussion of the basic principles of pharmacology, the science of drugs, as they relate to assessing the potential effectiveness and safety of an herbal supplement. Descriptions are provided of the components of the central nervous system that are dysfunctional in neurological and psychiatric disorders, and the targets of drugs used to treat these conditions. Taking all of these issues into consideration, a short checklist is provided to assist the potential consumer in determining, from a scientific standpoint, whether a particular product is likely to contain chemicals that beneficially affect brain function. The reader is encouraged to complete the first four chapters before proceeding to those describing individual plant products. The introductory chapters provide the context, concepts, and definitions essential for understanding fully the reasoning and conclusions drawn in the second part of the book.

Chapters 5 through 15 are devoted to a scientific assessment of the claims made for a select group of herbal products that are believed to have central nervous system effects. The pharmacological principles provided in the earlier chapters are applied in this analysis, with the checklist items used to guide the reader in the search for the truth. In this way, the reader can appreciate how answering a few key questions yields powerful insights into the potential benefit of these products.

The primary audience for this book is consumers interested in determining the value of herbal products purported to influence brain function. Others who will find this information of interest and value are students considering careers in the neurosciences or drug discovery, and scientists seeking an updated review of this field. By having the tools needed to make an objective and scientific assessment of these products, consumers are in a much better position to maximize the benefits of herbal supplements. This information will also make it possible to minimize the risks to one's health that comes with consuming these substances without adequate information on their effectiveness and safety.

# 1

## The Gifts of Eden

Adam wasn't hungry and was apprehensive about the potential consequences of eating the forbidden fruit. He was, however, convinced the plant material could provide benefits beyond its nutritional value. On the one hand, God told him that its consumption would be fatal, while the serpent contended the plant would impart new knowledge. Both were right. After eating the fruit Adam lost his home and immortality, and was made aware of the concepts of good and evil. He would need this new knowledge to survive in the world outside of Eden.

Besides its allegorical importance for Jews, Christians, and Muslims, this biblical account provides lessons for those interested in the therapeutic benefits of herbal supplements, also known as nutritional, dietary, or food supplements. Defined as a product that contains a vitamin, mineral, herb or other botanical, an amino acid, an extract, or any combination of these materials, the United States government considers dietary supplements to be foods rather than drugs. This has significant implications with regard to their regulation and the assurances provided to consumers. Because of this categorization, potential users must obtain on their own objective data about these products. The aim of this book is to provide such information.

The most fundamental question pertaining to dietary supplements is whether there is any evidence that they provide benefits beyond possible nutritional value. Written some 2,500 years ago, the Genesis account of Adam's introduction to these products indicates that humans have been familiar with the possible mystical and therapeutic powers of plants for quite some time. Moreover, the Old Testament account demonstrates that then, as now, there was uncertainty, and therefore risk, associated with the consumption of plants and plant products for religious, therapeutic, or, as in Adam's case, educational purposes.

The fruit consumed by Adam is unknown. In Old English, the word "apple" is simply a synonym for fruit. Regardless, when tempted to eat the plant product, Adam was at a distinct disadvantage to today's consumer. There was no historical record on its possible effects and no scientific data on its safety. Moreover, as the basic principles of pharmacology, the science of drugs, had not yet been established, he was unable to assess these properties himself. Rather, Adam had to rely solely on the word of others.

The constraints experienced by Adam remained for thousands of years until written records were maintained on the medicinal value of plants. More centuries passed before chemists were able to identify, and pharmacologists objectively study, the therapeutically active constituents in plant and animal products. Only during the past century has research revealed the diseases and disorders that are most responsive to these constituents, and to define precisely the appropriate doses to maximize safety and effectiveness in most individuals.

Anecdotal accounts about the potential benefits of dietary supplements have existed for thousands of years. Evidence includes pollen grains found on Neanderthal (Homo neanderthalensis) graves that were from plants lacking showy flowers, such as the yarrow (Achillea millefolium). It is inferred that these plants were placed there not for adornment, but to provide the departed a supply of medications in the afterlife.[1] This concept is based, in part, on the fact that many of the plants deposited on Neanderthal gravesites were subsequently described as therapeutics in early medical books, indicating that word of their therapeutic powers was passed on for millennia. For example, yarrow is mentioned in the Assyrian Herbal (800 BC), one of the oldest listings of therapeutically active plant products,[2] as well as in the Ebers papyrus (1500 BC) from Egypt. The Greek poet Homer described in *The Iliad* (800 BC) the use of yarrow to cure wounds, as did the Roman naturalist Pliny the Elder in his writings during the first century AD.[3]

A conservative estimate is that plants have been used as therapeutics at least since the appearance of modern man, some 200,000 years ago. It seems reasonable that as early humans foraged for food they would accidently discover the curative powers of some plants or take note of the fact that consumption of a certain type of seed, root, or fruit produced discernable effects on mood, sensory input, or alleviated general aches and pains. Indeed, as a species, humans are indebted to the many thousands of forgotten ancestors who became ill or died in the process of identifying plants and animals suitable for consumption. Thus, through trial and error, early man was able to identify plants that possess useful medicinal properties.

In addition to using plants to cure disease, they were also consumed in the ongoing quest for immortality. Recipes for "elixirs of life" were described in ancient writings. An example is the *Epic of Gilgamesh*, the story of a Sumarian hero that was recorded in 2000 BC.[4] After many travails, Gilgamesh obtained the plant of immortality from deep in the sea. Unfortunately for Gilgamesh, the plant was subsequently stolen by a serpent. This tale has many of the features of the biblical account of Adam and Eve. In the end, Gilgamesh returned home to Sumer to, like the rest of us, spend the remainder of his days as a mortal, awaiting the inevitable.

As in Genesis, ancient medical texts demonstrate that plant products have been used for therapeutic purposes for millennia. During most of this time no concerted effort was made to understand the reason for their effectiveness, or, in modern terminology, their mechanism of action. The first recorded attempts to synthesize therapeutics were made by European alchemists during the Middle Ages.[5] Besides their efforts to transform base metals into gold, the alchemists were interested in what made substances therapeutically useful as they wanted the power to transform basic materials into drugs. They were hindered in this quest, however, by the prevailing theories about the nature of matter and the causes of disease.

From the time of Aristotle to the seventeenth century, the use of plants in European medicine was based on the idea that all nature was composed of four basic elements: earth, air, fire, and water. Disease resulted from an imbalance of bodily humors. It was believed this imbalance could be countered by one or more of the four plant classes—cold, dry, hot, and wet—that corresponded to the four basic elements of nature. Mixtures of plants, usually from the same class, were preferred over a single specimen for treating medical conditions. For example, combinations of "cold" plants were used to treat fevers. Given these theories, drug discovery remained an empirical enterprise for thousands of years, with the identification of active plants and plant products left solely to chance.

By the seventeenth century, belief in the Aristotelian four elements was being challenged, most notably by the Irish chemist Robert Boyle.[6] Boyle understood that the precise identification and classification of the basic elements of nature were absolutely essential for understanding the universe, including drug actions. Thanks to his efforts, and those of many others, modern chemistry emerged in the nineteenth century. This made it possible to isolate, chemically define, and study the biological responses to plant constituents. As a result of these efforts, drugs were identified in plants that were first discovered by our distant ancestors. Many of these compounds, or their chemical derivatives, are still used today.

Given the historical records, and contemporary scientific data, there is no question that plants produce an abundance of substances that provide benefits beyond their nutritional value. However, not all plant constituents have been isolated and properly tested for effectiveness, and, unlike drugs, there is no government requirement that a manufacturer demonstrate effectiveness before marketing an herbal supplement. Like Adam, the consumer must rely on the word of others about the benefits of these products.

This book is designed to address this issue by providing basic information needed to assess the potential therapeutic value of plant products. Included are fundamental principles of pharmacology and about how drugs and natural products can affect various organs and organ systems. Explanations and examples are provided about what determines whether an ingested substance will find its way into the bloodstream, and then to the targeted site in the body at a concentration sufficient to have a beneficial effect. Other topics include the ways in which natural products may influence the blood levels of other substances, including drugs, and the likelihood that such interactions may diminish the effectiveness of prescription medications or alter normal body chemistry. While the principles described apply to all dietary supplements and drugs, emphasis is placed on factors that relate especially to herbal supplements purported to influence brain function. Individual chapters are devoted to a discussion of selected nutritional supplements that are said to enhance memory, or to aid in the treatment of depression, anxiety, insomnia, and alcoholism. These products were chosen because the promised benefits can be difficult to quantify and are more subject to influence by the power of persuasion than is the case with other therapeutics. This is why the use of such substances has been exploited over the centuries by shamans to maintain their social standing, and by charlatans for monetary gain. The properties of these products are described in the context of the basic principles of pharmacology and the results of scientific studies, both human and laboratory animal, aimed at determining effectiveness and mechanism of action. The approach taken in objectively evaluating these products can be used by the reader as a guide for assessing the information available on any dietary supplement. This work is intended for those who are curious about the potential benefits and risks associated with the use of food supplements. The information provided will be of particular value for individuals who, like Adam, are interested in how drugs and natural products affect us for good and evil.

# 2

## Transforming Plants into Gold

The notion that plant and animal products have therapeutic benefit is as old as mankind. This belief is based on solid empirical evidence accumulated by our ancestors over millennia. This discovery was undoubtedly the result of the need for early humans to forage for food. While consuming various roots, seeds, stems, and flowers, the ancients would occasionally stumble across something that alleviated a physical discomfort or altered their sensorium. Ultimately, the use of the plant for medicinal or ritualistic purposes became part of the culture. The validity of these observations was confirmed in the modern era with the isolation and study of pharmacologically active plant constituents. Indeed, the first drugs were purified substances obtained from ancient remedies or chemical derivatives of these plant products. It is easy to understand, therefore, why consumers are easily persuaded that a plant-derived dietary supplement may have therapeutic benefit. However, few of the products sold today were employed by early humans for the advertised purpose, and fewer still have undergone extensive scientific testing to demonstrate efficacy for their purported use. In many cases the active constituents of today's products are unknown, making it impossible to define their effects, both positive and negative, in humans.

An understanding of the historical development of therapeutics is useful for gaining a proper perspective on the value, and limitations, of plant extracts as remedies. Such information is particularly helpful when considering the claims made for such products.

## Prehistoric Evidence

The use by prehistoric humans of natural products as remedies is inferred from anthropological findings. For example, an analysis of pollen grains from a Neanderthal grave in Iraq revealed that the flowers were from six plants known today to have therapeutic properties. Because only a small fraction of plant life has medicinal value, it is probably no coincidence that these particular plants were placed on the grave. This finding is taken as evidence that Neanderthal, a humanoid species that became extinct 30,000 years ago, knew that these particular plants were of therapeutic benefit.[1] Pollen from these plants came from Centaurea solstitialis, Ephedra altissima, and species of Achillea, Senecio, Muscari, and Althea. It has been known for thousands of years that Centaurea (century plant),[2] Achillea (yarrow),[3] Senecio (groundsel),[4] and Muscari (grape hyacinth)[5] have wound-healing and antibacterial properties. Because of its mucilaginous properties, Althea (mallow) was used to protect irritated skin.[6] Ephedra contains ephedrine, a central nervous system stimulant known to impart a sense of well-being.[7] Presumably these plants were left graveside to provide the deceased with access to medications in the afterlife.

Other prehistoric evidence suggesting the use of plants for medicinal purposes includes the discovery of valerian root in caves inhabited 35,000 years ago by Cro-Magnon.[8] As the valerian root grows horizontally along the surface of the ground, it would be easily noticed and harvested for its nutrient and medicinal value.

## Early Documentation

Ancient literature contains numerous references to the therapeutic value of plants. Linneaus coined the genus name Achilla for yarrow after a description of its use in *The Iliad*, an eighth century BC epic poem on the Trojan War. In this narrative the hero Achilles is described as using this plant to staunch the bleeding wound of a fellow warrior during the battle for Troy. Given the hazards associated with Neanderthal life, it would not be surprising if they used this plant for the same purpose, long before the Greeks.

The Ebers papyrus is the oldest complete document describing medical remedies. The work is named after George Ebers, who purchased it in Egypt in 1873. The document is now in the University of Leipzig library. An English translation of the Egyptian hieroglyphics is available.[9] The Ebers papyrus is thought to date to about 1500 BC, making it one of the oldest sources of written information on how our ancestors treated injuries and disease. The remedies in the Ebers papyrus contain mixtures of animal, vegetable, and mineral materials. Animal constituents include stag's horn and dried dung. Mineral components mentioned are natron, a form of sodium carbonate found in the Egyptian desert, and malachite. Myrrh, frankincense, coriander, mustard seed, and cumin are some of the plant products recorded in the Ebers papyrus. Before use, these materials were normally dried, ground, and added to honey if they were to be taken orally, or mixed with oil or fat if they were to be applied topically as a poultice. Some ingredients, such as natron and frankincense, were used for a variety of ailments. Because the same ingredient combinations were employed for treating different conditions it is assumed their use was often more related to tradition rather than to established effectiveness. As an example, in the fifth century BC Theophrastus described megaleion, a poultice he claimed relieved the inflammation resulting from a wound.[10] The ingredients of megaleion, all of which are mentioned in the Ebers papyrus, are burnt resin, oil of balanos, cassia, cinnamon, and myrrh. Myrrh, a gum resin from Commiphora myrrha, is now known to contain antibacterial and antifungal compounds that do, in fact, improve wound healing.[11] This indicates that the early use of some mixtures, such as megaleion, was evidence-based, not just driven by habit, availability, or religious beliefs. Contemporary research indicates, however, that not all of the ingredients contained in early remedies were of medicinal value if used as recommended. In fact, many may have been toxic. Overall, the use of these products was driven primarily by tradition or folklore. For some, this may still be the case today.

The value placed by the ancients on medicinal plants is illustrated by Queen Hatshepsut's quest for myrrh. Queen Hatshepsut was pharaoh of Egypt for over two decades, beginning in 1479 BC. History records that one of her accomplishments was to bring myrrh trees to Egypt. Myrrh trees were native to the Land of Punt, which is now part of Ethiopia.[12] The priest-physicians of Egypt needed myrrh as incense and for medicinal purposes. Queen Hatshepsut sent five Egyptian warships across the Red Sea to Punt. The ships were loaded with gifts, presumably to exchange for live myrrh trees. These trees were planted in Egypt, where their descendants can still be found to this day.

Another early listing of remedies and their uses is contained in the Assyrian Herbal.[13] Dating from the seventh century BC, these 660 cuneiform clay tablets were preserved for posterity by being baked during the burning of Ashur, the ancient religious capital of Assyria, a region of contemporary Iraq. More than 300 plants are listed in the Assyrian Herbal, some of which are still used as therapeutics, including myrrh and opium. One plant family, the Solanaceae, is particularly notable in that several genera produce chemicals with marked effects on central nervous system function. These actions were recognized by the ancients because ingestion causes sedation, disorientation, and coma, depending on the amount taken. Such effects are readily apparent to the consumer and to those who witness its use. Chemicals first identified among plants of the Solanaceae family, or derivatives of them, are still used today. Most notable are the belladonna alkaloids atropine, hyoscyamine, and scopolamine. The Solanaceae represent only a few of the plants employed by the ancients for modifying central nervous system function (see Table 2.1).

**Table 2.1** Selected Examples of Plants Used Historically for Modifying Central Nervous System Activity

| Date | Plant | Active Ingredient* | Reported Effect |
|---|---|---|---|
| 40,000 BC | Ephedra | Ephedrine | Stimulant[1] |
| 20,000 BC | Valeriana | Valerenic acid | Sedative[8] |
| 3000 BC | Camellia sinensis | Caffeine | Stimulant[4] |
| 1550 BC | Papaver somniferum | Morphine | Analgesic/Sedative[9] |
| 800 BC | Cannabis sativa | Tetrahydrocannabinol | Antidepressant[13] |
| 100 AD | Hyoscyamus | Hyoscyamine | Sedative[16] |
| 400 AD | Peonia | Paeoniflorin | Analgesic[18] |
| 1550 AD | Mandragora | Atropine | Sedative[23] |
| 1600 AD | Melissa officinalis | Rosmarinic acid | Antidepressant[15] |

*In all cases, the active ingredient was not identified until the nineteenth or twentieth centuries.

Some of these, such as the sedatives, analgesics, and stimulants, were easy to identify by the user. For such plant products there is little need for sophisticated clinical or laboratory animal experiments to prove they have a biological effect. In contrast, a purported beneficial effect on mood, such as that reported for Melissa officinalis, is much more difficult to demonstrate, even with clinical studies, given the challenges associated with quantifying the symptoms of depression.

## Western Culture

There is a rich historical record of therapeutics used in the Greco-Roman era. Much of what is known comes from the writings attributed to Hippocrates, a fifth century BC Greek physician. Hippocrates and his followers developed the four humors theory to explain the cause of disease. The four humors of Hippocrates were blood, phlegm, yellow bile, and black bile. He theorized that these substances were in balance in a healthy individual, and that illness resulted when the relationship among them was disturbed. Medical treatment, according to Hippocrates, should be aimed at restoring this balance. The biological humors corresponded to what were considered by Aristotle the four basic elements of nature: fire (blood), water (phlegm), air (yellow bile), and earth (black bile). It was also reasoned that they related to the climates associated with the four seasons of the year: summer and hot (blood), spring and wet (phlegm), fall and dry (yellow bile), and winter and cold (black bile). This association with the weather was coupled further with four human temperaments: sanguine, phlegmatic, choleric, and melancholic, respectively.[14] Because of these purported relationships, plant remedies used during the Hippocratic period were classified as warming, moistening, drying, or cooling.

The Hippocratic tradition of medicine was followed by physicians for centuries. This is illustrated by the fact that in the sixteenth century John Gerard compiled a list of contemporary plants used as remedies, classifying each as warming, moistening, drying, or cooling.[15] He then subdivided them further into as many as four additional levels of effectiveness. For example, Gerard considered dill seed (Anethum graveolens) hot in the 2nd degree and dry in the 1st degree, while fennel (Foeniculum vulgare) was hot and dry in the 3rd degree. These distinctions were made on the basis of the odor and taste of the plant product, not the therapeutic effectiveness or use. In fact, no matter how elaborate the classification system based on these properties, none was very effective for predicting which plants would be best for treating a particular condition in an individual patient.

Although the Hippocratic humoral theory of disease and its treatment was dominant for centuries, not all remedies employed during that time fit within this classification system. An example was mithridatum, a remedy devised in the first century BC by Mithridates, King of Pontus, a region in northeastern Turkey, and his associate, Crataeus, a botanist and physician. The composition of mithridatum and the history of its use were recorded by Celsus, a Roman who lived during the first century AD.[16] Celsus described several theories on the cause of disease. In his work *De Medicina*, he stated that the treatment of medical problems should be based on evident causes. Because Celsus demanded proof, his position is a direct challenge to the idea of humors and other theoretical constructs used to explain the cause of disease and to classify treatments. Mithridatum, which is a mixture of more than 30 plants, was recommended by Celsus for treating wounds and bruises, and for alleviating the suffering associated with falls. Mithridatum remained a popular remedy in

Europe until the nineteenth century. These 2,000 years of use is not surprising in that it is now known that some of the plant products in mithridatum display anti-inflammatory and antioxidant properties. This explains its utility in treating many of the conditions described by Celsus.[17] Thus, the writings of Celsus, and the use of mithridatum, called into question Hippocrates' unifying theory of humors.

Around 400 AD the Roman Psuedo-Apuleius compiled a list of 130 herbal remedies.[18] From a twenty-first century perspective, the medical value of these herbs ranges from useless to undoubted effectiveness. An example of the former is the magical properties attributed to ironwort (Sideritis heraclea). Psuedo-Apuleius claimed that a person carrying a branch of this plant became invisible and was therefore protected from robbers. This account contrasts with his accurate reporting that skin lesions and irritations were effectively treated with apolinaris (Hyoscyamus), a plant that was subsequently found to have local anesthetic properties. He also recommended that squill (Urgenia maritima) bulbs be used to treat dropsy, a symptom of heart failure. It is now known that squill produces compounds that are, in fact, effective in treating this condition. The written account of Pseudo-Apuleius is also historically important because he failed to mention the humoral theory of disease, and he described only the use of single plants, excluding mention of plant mixtures. This suggests erosion of the Hippocratic approach even at this early time and a shift in thought to emphasize the value of individual herbs as treatments for specific conditions. Nonetheless, in Europe the Hippocratic theory of disease and its treatment predominated through the Middle Ages.

## Alchemy

The demise of the Hippocratic theory of disease can be traced to the rise of alchemy in the thirteenth century AD. Alchemy, which began as a philosophy, was practiced in Egypt, Babylonia, India, and China prior to its arrival in Europe in the late Middle Ages. Early alchemists were purported to possess magical powers, including the ability to transform base metals, such as iron, into noble metals, such as gold or silver. It is thought that alchemy evolved as humans noticed that the physical state of matter could be altered. Thus, water turns to ice on freezing, and to steam on boiling, and wood becomes ash upon burning. It was known from prehistory that fermentation converted plants into sedating beverages, beers, and wines, which were widely used for cultural and medicinal purposes. Alchemists were the first to use experimental procedures systematically as they attempted to understand the transformation of matter. To this end they developed and employed techniques such as distillation, calcination, which is formation of a salt, and sublimation, the removal of volatile compounds by heating, that are still routinely used by chemists. It is believed that distilled spirits, which have a higher alcohol content than fermented beverages, were developed by Greek alchemists in the first century AD. Alchemists are also thought to be responsible for devising methods to separate metals from ores and for producing cosmetics.

Records dating back to 300 AD describe Greco-Egyptian alchemical recipes for preparing metal alloys.[19] European alchemists drew heavily from the Arabs in developing the discipline. Rhazes (860-932), a Persian physician who first described smallpox and measles as distinct disorders, was also an alchemist interested in creating therapeutics.[20] With the spread of alchemy throughout Europe, the number of practitioners increased, along with information on the chemical nature of matter. The work of the alchemists made it apparent that the old Aristotelian concept of four basic elements and the Hippocratic humoral theory of disease were inaccurate and misleading. Certainly alchemy rendered them obsolete.

An example of the use of alchemical techniques to improve medical treatment is provided in a manuscript by Giovanni Andrea (died 1562), a Jesuatti friar from Lucca, Italy.[21] Friar Andrea detailed recipes for the remedies prepared by members of his religious order over the previous 200 years. Among these concoctions was an elixir of life prepared alchemically by slowly heating a mixture containing some 50 plants and animal products, including spices, fragrant herbs, nuts, dried fruits, and honey. The resultant distillate was mixed with an equal weight of brandy and then redistilled. Friar Andrea considered this second distillate "more valuable than gold." He wrote that an 80-year-old would, within a month, take on the appearance of a 40-year-old if he drank a specified quantity of this elixir each day.[22]

Paracelsus (1493-1541) was a physician trained in alchemy who hoped to exploit the therapeutic potential of natural products. He believed that every vegetable and mineral product contained a unique essence that was responsible for its biological activity. Using the tools of the alchemist, Paracelsus set out to extract this essence and thereby improve the efficacy of the remedy. Rather than accepting the notion of four humors and four elements, he believed, like most alchemists, that all materials were composed of three basic principles: sulfurs, mercuries, and salts.[23] Sulfurs were thought to be the combustible part of a substance, while mercuries were the smoke or gas emitted when the material was burned, and salts the ash that remained after combustion. While it is known today that this concept is no more accurate than those put forth by Aristotle and Hippocrates, the work of Paracelsus and other alchemists was instrumental in discrediting the earlier theories. This was a critical step in the development of modern chemistry and the ultimate identification and characterization of plant essences, as originally conceived by Paracelsus.[24,25] While alchemists were never able to transform base metals into gold, their work provided the foundation for the modern pharmaceutical industry and the development of medications that have eased the suffering of millions. To paraphrase Friar Andrea, their work led the way in making it possible to transform plants into gold.

## Chemistry

In his 1661 publication *The Sceptical Chymist*, Robert Boyle proposed that matter is a collection of elements that can be arranged in various ways to yield different chemical substances.[26] Boyle developed this idea by inventing a series of dialogues between individuals expressing support for his experimentally based concept of numerous basic elements for matter, and skeptics who continued to believe that all matter was composed of either three (Paracelsian theory) or four (Aristotelian theory) elements. Boyle's insights were the beginning of modern chemistry and drug discovery.

The concept of chemistry as an experimental science was proposed by Francis Bacon (1561-1626) a few decades before Boyle's *The Sceptical Chymist*. In his 1620 treatise titled *Novum Organum*, or *New Method*, Bacon proposed that scientific knowledge is accumulated through experimentation aimed at testing specific hypotheses.[24] This idea remains the foundation of modern science. While perhaps apocryphal, it is said that Bacon conceived of the experimental method after observing that a chicken frozen in a snow bank did not decompose. He then hypothesized that frozen meat could be safely stored for long periods, and therefore consumed over time. By designing and executing experiments to test this theory, Bacon is attributed with demonstrating conclusively the value of freezing for storing perishable foods.

As the seventeenth century ended, a search began for the basic elements of nature as conceived by Boyle. Soon it was discovered that all living matter consists of carbon, oxygen, hydrogen, and nitrogen. The discovery of other elements followed rapidly. By the mid-nineteenth century more than 60 atoms had been identified. The utility of this information wasn't fully realized until the Russian chemist, Dimitri Mendeleev, described the relationships among them, a puzzle that he and others had worked on for years.[27] While various systems were proposed, none fulfilled the criteria needed for a useful classification system. Mendeleev reported that the solution appeared to him in a dream. Upon awakening, he quickly sketched a chart aligning the various elements, or atoms, into categories based on their known composition and physical properties. His inspiration resulted in the creation of the Periodic Table of Elements, or periodic chart. As this classification scheme made it possible to predict chemical interactions, it set the stage for the selective synthesis of specific chemical agents, and for identifying the atomic structure of all elements in nature, including the constituents of plant extracts. The accuracy of Mendeleev's insight is proven by the fact that his original Periodic Table of Elements contained blank spaces for atoms he predicted existed but that weren't yet discovered. In subsequent years all of these elements, as well as others, were identified. In 1955, 48 years after his death, element number 101 was found and, appropriately, named mendeleium in his honor.

The Periodic Table of Elements effectively ended the practice of alchemy and initiated the era of modern chemistry. It was now possible to isolate and chemically identify biologically active plant constituents and to synthesize chemical derivatives of them. This gave scientists the tools to fulfill Paracelsus' dream of purifying the plant essence and, as he envisioned 300 years earlier, to modify it for therapeutic gain.

# 3

## Thinking Like a Pharmacologist

The United States Pure Food and Drug Act became law on June 30, 1906. This legislation mandated the federal inspection of meats and banned the sale of adulterated food products and poisonous drugs. In 1938 the Federal Food, Drug, and Cosmetic Act established the United States Food and Drug Administration (FDA). This agency was given the responsibility for ensuring the safety of food, pharmaceuticals, and cosmetics and, subsequently, was empowered to assess and verify drug efficacy. Given these and other legislative mandates, a drug candidate must now undergo years of rigorous laboratory animal and human testing before the manufacturer gains FDA approval for sale to the general public. Hundreds of millions of dollars are needed to cover the costs associated with meeting FDA criteria for a single, new drug product. Among these requirements is detailed information on its chemical and pharmacological properties. Convincing evidence must be submitted that the compound is safe and efficacious at the recommended dose. Side effects and toxicities must be identified, including information on potential interactions with other drugs that may be taken by the patient. The intended use must be defined precisely on the basis of clinical research results. Data are required on the purity and stability of the manufactured product. After a drug is approved, federal inspectors routinely monitor its production and use to ensure product consistency and appropriate marketing. Given these safeguards, it is not necessary for consumers to be concerned about the safety and effectiveness of prescription or over-the-counter medications. While not all drugs are effective in all patients, and there can be idiosyncratic responses, consumers can reasonably assume that if an FDA approved product is taken as directed it will likely display some efficacy as a treatment for their condition and that they will be alerted to potential side effects. Accurate information on the limitations of use and possible toxic reactions is usually obtained from the health care provider and is readily available in the manufacturer's description of the product and in other forums.

The situation is different for herbal supplements. Given their growing popularity in the 1980s, Congress needed to decide whether these products should be subject to federal regulations covering foods or drugs. In the United States this resulted in the Dietary Supplement Health and Education Act of 1994. As this legislation defined these products as food supplements, producers are not required to provide proof of safety or any health benefit before offering them for sale. While a statement regarding safety must be submitted to federal authorities, the burden of proof is on the government to raise questions about possible dangers associated with use. If the government registers no objection within a specified period, the product may be sold. Given the lack of regulation and oversight, those selling such products are forbidden to refer to them as drugs and to advertise any therapeutic benefit. To comply with this requirement, product labels often contain a disclaimer indicating it is not to be used to diagnose, treat, cure, or prevent any disease. As a practical matter, however, broad, misleading claims about the therapeutic benefits of dietary supplements are commonly encountered, especially on the Web.

The number and popularity of these products has continued to expand, with more than $5 billion in sales in the United States in 2009. Because of growing concerns about product contaminations and the quality of the ingredients used to manufacture some supplements, new regulations were enacted in 2003 to allow government inspectors access to company manufacturing records to monitor quality control. However, it cannot be assumed that all herbal product manufacturers routinely undergo such inspection given the number of companies, the fact that many are located abroad, and the shortage of FDA personnel. The absence of federal requirements regarding safety and efficacy, and the minimal manufacturing oversight, leaves the consumer responsible for assessing the potential risks and benefits of these products. Besides a paucity of objective research data, this determination is complicated by the fact that neither the potential user nor the manufacturer can always be sure of the number and type of biologically active constituents in these extracts. This makes it difficult to assess what effect the supplement may have on otherwise healthy individuals, let alone those with a chronic or acute illness, those taking prescription medications, or who are undergoing other kinds of treatments, such as radiation therapy.

When developing new drugs, safety and efficacy issues are addressed by basic and clinical pharmacologists. To this end, pharmacologists consider a number key principles. While this task is simplified when examining a single, purified substance, these same principles can be applied, although to a more limited extent, to plant extracts when deciding whether a product might be of benefit. The information contained in this chapter is intended to provide a broad overview of these basic pharmacologic principles and to define terms and concepts useful in making an informed judgment about the potential utility of an herbal product. The aim is to help the consumer think like a pharmacologist when considering the possible use of a dietary supplement. Although the quality and quantity of the publicly available research data are limited for these products, an understanding of these principles and terms will enable the consumer to make a more informed decision about the potential value of a particular product.

## The Origins of Pharmacology

Pharmacology, the study of drug actions, was not recognized as an independent scientific discipline until the mid-nineteenth century. Up to that time pharmacological principles were included in medical courses on Materia Medica, which dealt with the origin, preparation, and administration of therapeutic agents. As most of these therapies were mixtures of plant extracts, little was known about the active constituents. Rather, Materia Medica focused mainly on the preparation and clinical use of herbal products discovered empirically in earlier centuries. Little effort was made to search systematically for new therapeutics or to understand the mechanisms of action of these products.

This situation changed dramatically with Dmitri Mendeleev's development in 1869 of the Periodic Table of Chemical Elements and the advent of modern chemistry. By classifying the atomic constituents of matter, Mendeleev settled the centuries-old dispute between Aristotelians and alchemists about the number of essential elements in nature. Once the atomic basis of matter was understood, it became possible to identify and manipulate chemical structures. Mendeleev's accomplishment made it possible for chemists to characterize precisely the essence of therapeutic plant extracts, an objective initially, but unsuccessfully, pursued by Paracelsus some 300 years earlier.

An early example of this new approach to understanding drug action, and the response to plant extracts, was the work of Friedrich Serturner, a nineteenth century German chemist. Serturner was interested in identifying the active constituent of opium, a gum resin from the poppy Papaver somniferum. Opium was often used by Paracelsus in his remedies, which often contained another gum resin, laudanum, an extract from Cistus laudanifer. Over the years the mixture of these two plant extracts came to be called laudanum, even though, as we know now, the clinical benefit is derived primarily from the opium components. Today, laudanum, which is prescribed for its analgesic properties and for controlling diarrhea, is known as Tincture of Opium. As Serturner knew that the opium extract contained many ingredients, he set out to separate chemically the various agents in an attempt to determine whether the beneficial effects were mediated by one or more of them. Because opium is known to cause drowsiness and sleep, he named one of these purified compounds morphine, after Morpheus, the Greek god of dreams. Serturner's effort is the first recorded purification of an active constituent from a plant extract. Eventually the chemical structure of morphine was identified, making it possible to synthesize it and hundreds of structurally related agents, some of which are used as drugs today.[1,2]

Another example of this approach was the purification of salicin. Since at least 400 BC the extract of the willow tree (Salix spp.) was used to treat a variety of conditions. In the sixteenth century Friar Andrea described how this extract was employed by the Jesuatti Friars for the treatment of gout. This remedy was prepared by holding a section of green wood from a willow tree over a fire and collecting the foam that appeared on the unheated end of the stick. Andrea reported that topical application of this foam to a painful, gouty joint "will soon lift the pain."[3] In the late nineteenth century salicin, the chemical precursor of what proved to be the active constituent in humans, was isolated from the willow extract. It was then found that in the body salicin is converted to salicylic acid, an anti-inflammatory agent. Subsequently, chemists synthesized acetylsalicylic acid (aspirin), a chemical derivative of salicylic acid that could be taken orally. To this day aspirin remains one of the most popular drugs for treating some forms of pain and inflammation. Salicylic acid was first prepared from salicin in 1838, and aspirin first synthesized in 1899.[4]

A common feature of these two examples is that both opium and willow bark extract were employed for centuries by various cultures for defined medicinal properties. This history of use greatly increased the likelihood that these preparations contained pharmacologically active substances, making them logical choices for purification.

The science of pharmacology evolved rapidly once it was possible to isolate and identify clinically active constituents in plant products, and to synthesize safer and more effective derivatives. It was appreciated that to exploit fully this new approach to drug discovery, it was necessary to define the manner in which these chemical substances, now referred to as drugs, act on the body. This is the task of the pharmacologist.

## Pharmacodynamics

Pharmacology is divided broadly into two subdisciplines: pharmacodynamics and pharmacokinetics. Pharmacodynamics relates to the drug mechanism of action, which is the way the drug affects the body. As an example, pharmacodynamic studies are those aimed at determining precisely how morphine relieves pain, depresses central nervous system function, and causes nausea and constipation. Such information is critical in designing and synthesizing more effective drugs with fewer side effects. Pharmacokinetic studies are those concerned with understanding the way the body affects the drug. These include defining the absorption, metabolism, and excretion characteristics of the chemical agent. Such work is important in ensuring the drug reaches the desired site of action and remains there for a sufficient period of time, and at the necessary concentration, to have the intended effect.

As for pharmacodynamics, it is now appreciated that most drugs act by attaching to specific cellular components to activate or inhibit biochemical functions. Typically the attachment, or binding, site for most contemporary drugs is a transmitter or hormone receptor, or an enzyme that regulates some component of cellular function in the patient or an invading organism. These types of targets are particularly common for drugs used to treat central nervous system disorders.

The idea that drug actions are mediated through attachment to specific receptors was first proposed in 1905 by John Langley, professor of Physiology at Cambridge University. His theory was related, in part, to work performed a few years earlier by Paul Ehrlich in Germany.[5,6] Ehrlich studied the mechanisms responsible for the development of immunity to infection. His work led to the development of a diphtheria antitoxin that was used immediately to quell an epidemic of this disease. As a result of his studies, Ehrlich proposed that antibodies are produced in response to the selective binding of bacterial toxins to certain cells.[5] As these toxin binding sites are specific for chemically distinct foreign substances, Langley subsequently reasoned that such a mechanism may be responsible for the actions of drugs as well. He called the theoretical drug attachments site the "receptive substance." The Langley hypothesis met considerable resistance from the scientific community. Nearly 30 years of research were needed to convince others that drugs act by attaching to specific sites in tissue.

Langley's theory was based in large measure on his studies defining the actions of natural products, such as atropine and curare, on nervous system function and physiology.[7] His work demonstrated that these two, chemically distinct, substances block the normal physiological response to a single transmitter, acetylcholine, which is released by certain nerve endings. Whereas atropine and related alkaloids derived from Atropa belladonna block the effect of acetylcholine on some nerves, they have little effect on the ability of this transmitter to induce skeletal muscle contraction. In contrast, curare, which is extracted from plants in the genus Strychnos in the Logianaceae family, selectively blocks the effect of acetylcholine on skeletal muscle but not on nerves. This ability to differentiate the physiological actions of acetylcholine by applying chemically distinct substances suggested to Langley that these agents interact in a selective manner with specific and distinct sites on nerves, in the case of atropine, or skeletal muscle, in the case of curare. He referred to these sites as receptors, with one group being sensitive to atropine and the other curare. Subsequently, these two classes of acetylcholine receptors were classified as muscarinic and nicotinic, with the former being those most sensitive to atropine and the latter to curare.

Other data supporting the drug receptor concept included the discovery that chemical isomers differ pharmacologically. Isomers are compounds with the same molecular but different structural formula. In other words, while the atoms in chemical isomers are identical and are linked in the same way, one or more of them is oriented differently in space. Langley noted that most of the pharmacological response to hyoscyamine, a constituent of Hyoscymus niger, is mediated by the l-isomer, with the d-isomer being virtually inactive. He concluded from this that only a cellular component with a high degree of selectivity for chemical structures, such as a receptor site, could make such a subtle distinction between two nearly identical agents.[7] This result would not be expected if drugs acted simply by nonselectively modifying cell function.

It was also noted by Langley and others that drugs induce their effects by either activating or inhibiting receptors. Compounds that mimic the endogenous transmitter or hormone by stimulating a receptor are known as agonists, while those that inhibit the action of the endogenous agent are known as antagonists. Although antagonists bind to the receptor, they are incapable of generating a response. Their attachment instead prevents the endogenous agent from acting, thereby diminishing, or totally blocking, the effect of the hormone or neurotransmitter that normally interacts with the site.

The controversy engendered by Langley's receptor theory was ultimately resolved in his favor. The factors supporting receptors as the site of drug action were summarized by A. J. Clark, professor of Pharmacology at the University of Edinburgh, in a 1933 book titled *The Mode of Action of Drugs on Cells*.[8] Among the pharmacological principles developed by Clark was the importance of dose-response experiments in understanding drug action. Such analyses are especially meaningful when it was possible to study purified chemical agents. Using these data Clark showed that the laws of physical chemistry could be applied to explain the interactions of drugs with specific, chemically selective sites, or receptors. His work, which yielded a mathematical framework for explaining pharmacological data, led to wide acceptance of Langely's receptor hypothesis.

The receptor theory of drug action was unequivocally confirmed in the 1970s and 1980s, first with the biochemical identification and characterization of these sites and, subsequently, with their isolation and purification, and the cloning of the genes responsible for their production.[9,10] Receptors are now among the more important targets for drug development, especially for agents intended to treat neurological and psychiatric disorders. Unless proven otherwise, it is now assumed that any chemical that affects central nervous system function interacts with a specific site in the brain. Candidate sites include transmitter receptors or transporters, or the enzymes responsible for the production or metabolism of neurotransmitters and other endogenous chemical mediators.

## Pharmacokinetics

For an agent to be therapeutically useful it must first reach its site of action in the body and remain there at an appropriate concentration and for a period of time sufficient to induce a clinically meaningful effect. This is not a trivial matter as animals have evolved a number of ways to prevent foreign substances from gaining entry to the body and to dispose readily of such agents if they appear in the bloodstream. More lipophilic ("lipid loving") compounds, those that are more soluble in fat, are more rapidly and completely absorbed by the body than those that are more hydrophilic ("water loving"), or more soluble in water. Bioavailability is a measure of the extent to which an agent is absorbed into the bloodstream. For example, for a drug with a 50% oral bioavailability, one-half of the ingested dose will find its way into the systemic circulation and possibly be available for inducing a pharmacological effect.

Even if an agent is sufficiently lipophilic to cross from the gastrointestinal tract into the general circulation after oral administration, there is no guarantee it will gain entry into the brain. As the brain vasculature has tighter junctions than blood vessels in other organs, only the most lipid soluble agents are able to diffuse across this blood-brain barrier and penetrate into the central nervous system. Drug diffusion is a passive process whereby the chemical crosses tissue membranes after dissolving in them, just as rainwater soaks the skin after passively diffusing across clothing. In some cases, more hydrophilic chemicals are transported across tissue membranes by specialized proteins in the gastrointestinal system and other organs, including the brain. This mode of absorption is an active process requiring energy. The transporter is a protein located on the cell membrane to which compounds with the appropriate chemical structure attach. The transporter then transfers the agent into, or out of, the cell just as a tow truck moves a car from one side of the street to the other. These transporters evolved to ensure that certain essential, but hydrophilic, nutrients, such as some amino acids and glucose, are taken into tissues even though they do not diffuse across cell structures. Drugs and toxins, therefore, are able to reach the brain following oral administration if they are sufficiently lipophilic to penetrate the tissue by passive diffusion or their chemical structures are similar to the endogenous substrates for transporters.

Once a chemical reaches its target, a sufficient quantity must accumulate to activate or inhibit the site. This is a function of the dose administered, the extent to which the compound is absorbed, and its affinity, or attraction, for the target. Thus, while an agent may be capable of reaching the systemic circulation and penetrating into the brain, to obtain a measurable response it may be necessary to administer large quantities if its absorption is limited or its affinity is low.

To be useful clinically, a drug must be active for several hours following its administration. The duration of drug action is directly related to the amount of time the agent is located at the target in sufficient quantities to be active. One measure of this is the drug half-life, or $t_{1/2}$. The half-life is the amount of time it takes for the drug concentration in the blood to be reduced by half. In general, the longer the half-life, the more prolonged the duration of action. As the half-life is directly related to the rate of metabolism and excretion, these are critical variables in determining the response to a drug. Hydrophilic agents are often rapidly excreted in feces or urine without undergoing a chemical change. More lipophilic substances, however, are generally metabolized by specialized enzymes located throughout the body but highly concentrated in the liver. The cytochrome P450 family is one of the major classes of enzymes responsible for the metabolism of many drugs and endogenous substances.

The characterization of pharmacokinetics is critical for evaluating whether a particular chemical has therapeutic potential. For example, if studies indicate that the test agent is not absorbed into the blood following ingestion, it is unlikely to be a drug candidate, at least with this route of administration. Likewise, if it is found that the test substance, although absorbed, is rapidly metabolized and cleared from the systemic circulation, it is a poor drug candidate as it will be difficult to achieve and maintain the levels at the target site necessary for the intended response.

Pharmacokinetic studies are also crucial for appropriately using drugs already known to be effective. In these cases it is important to know whether the agent is absorbed primarily by passive diffusion or by a transporter. If the latter, there is a potential for adverse drug interactions if the agent is administered along with another compound that utilizes the same transporter for absorption. In this case the two substances will compete for the limited number of transporter proteins, resulting in a decline in the absorption of one or the other, or both. Similarly, a drug that is metabolized by a particular cytochrome P450 enzyme may, if present in sufficient concentrations, slow the metabolism of other agents, both drugs and endogeneous substances. Moreover, some drugs are known to stimulate the production of cytochrome P450 enzymes, increasing the rate of their own metabolism and the breakdown of any other substrate for this enzyme. Such effects on drug metabolizing enzymes can lead to a dramatic increase or decrease in the blood levels of other substances, resulting in side effects or toxicities, or a decrease in effectiveness, which would not be encountered otherwise. As is evident, information such as this is important in evaluating the potential utility, and safety, of herbal supplements as it is known that some of them affect the pharmacokinetics, and therefore the safety and effectiveness, of prescription drugs.

## In Vitro and In Vivo Studies

In vitro and in vivo experiments are routinely performed for pharmacodynamic and pharmacokinetic studies. In vitro experiments are those conducted on organs or tissues after their removal from the animal. Such studies are useful because they eliminate or minimize absorption or metabolism as variables, making it possible to examine more directly the pharmacodynamic effect of the drug candidate, to identify the enzymes responsible for its metabolism, or to characterize the transporter needed for absorption or elimination. In vivo experiments are those conducted in animals to study the overall effects of a test substance on responses that are best measured in intact organisms. These include, for example, changes in behavior, blood pressure, and the response to painful stimuli. Because the experimental conditions for in vitro studies are artificial, it is sometimes difficult to extrapolate the results of such experiments to what might occur in the animal following systemic administration of the test agent. A drawback for both in vitro and in vivo laboratory animal studies is that they typically employ nonhuman subjects or tissues. As it is known that certain human receptors differ from those in other species, and that pharmacokinetic properties may vary markedly between humans and other animals, care must be taken when making predictions about the clinical potential of test substances solely on the basis of in vitro and in vivo laboratory animal experiments.

## Pharmacology and Herbal Supplements

Because most herbal dietary supplements are plant extracts, they contain scores, if not hundreds, of chemical substances. Because the constituents in these extracts can vary considerably from batch to batch, both qualitatively and quantitatively, it is difficult, if not impossible, to conduct the types of pharmacodynamic and pharmacokinetic studies needed to determine whether these products possess the pharmacological characteristics needed for therapeutic activity. The variation in extract composition is due to numerous factors. Included are the age of the plant, the environment in which it is grown, the portion of the plant used to prepare the extract, and the extraction procedure. A pharmacological assessment of herbal dietary supplements is particularly challenging for products taken to modify central nervous system activity. This is due in large measure to the difficulties associated with assessing whether an active constituent penetrates into the brain, and the lack of objective measures for some behavioral responses.

A detailed and precise pharmacological analysis is best conducted on a single, purified substance. As long as the active constituent of a dietary supplement is unknown, it is impossible for a pharmacologist to draw firm conclusions about the likelihood of the product having clinical benefit. Although there is evidence suggesting the identity of the active component of some herbal dietary supplements, and pharmacological studies conducted on them, such information is generally lacking for most of these products. Even when such studies have been performed, significant questions remain about the relevance of the pharmacological data as long as there is uncertainty about the identity of the active component. Add to this the limitations associated with understanding the relationship between in vitro and in vivo laboratory animal studies and human responses, and it is not surprising that significant skepticism remains among members of the scientific community and lay public about the health benefits of many herbal supplements.

## Herbal Supplement Pharmacology Checklist

Because of the importance of defining pharmacological properties in assessing the clinical potential of a new drug agent, they should be considered when contemplating the use of a dietary supplement. Whereas all of the essential pharmacological data for prescription and over-the-counter drugs are documented and have been reviewed objectively by government scientists and agencies, this is not the case for herbal products. This poses a challenge for consumers as most are unaware of the kinds of questions asked by a pharmacologist when examining the therapeutic potential of a drug. Accordingly, provided below is a checklist of items to consider and issues to investigate when attempting to determine the likelihood that the potential benefit attributed to a particular herbal supplement will outweigh the risks associated with its use.

1. **History of use.** For many herbal products this will be the most objective information available. Usually it is readily obtained online and in published literature. If it has been reported for centuries that a particular plant extract has a certain therapeutic benefit, there is a high probability it will display some efficacy in this regard, assuming the commercial product is prepared in a manner similar to that utilized by the ancients. It is important to be alert, however, to claims of a new use for a plant extract that is generally accepted to be effective for other conditions. For example, claims that an extract known to display anti-inflammatory activity has been discovered to display antidepressant or anticancer activity should be met with skepticism. Strong scientific evidence of efficacy for these other uses must be provided because, unlike the established anti-inflammatory action, they have not yet withstood the test of time.
2. **Absorption.** An effort should be made to determine whether there are any data on the bioavailability of substances contained in the plant product after it is taken by the recommended route of administration. As such experiments are relatively easy to perform, lack of such information should raise suspicion about whether there is any appreciable absorption into the systemic circulation. No agent or extract will have a direct effect on the body, except perhaps for the gastrointestinal system, if it is taken orally but fails to get into the bloodstream. Although the active constituent of the extract may be unknown, some assurance should be given, based on experimental data, that components of the product are detectable in blood following administration of the product. Bear in mind that many known plant products are very hydrophilic and therefore poorly absorbed following oral administration.
3. **Distribution.** It is valuable to know whether studies were performed demonstrating that the extract constituents appear in the target organ. For products reported to affect central

nervous system activity, such as those recommended for treating anxiety or depression, it would be reassuring to find laboratory animal data showing the appearance of constituents in the brain following the route of administration recommended for humans. The absence of such data should not be taken as conclusive evidence that an active substance does not get into the brain. It is always possible that the effective component of a plant extract has not yet been identified. Moreover, as noted earlier with salicin, there is the possibility that the active compound is a metabolite of one of the extract components, rather than any of the herbal constituents themselves.

4. **Duration of action.** If blood levels have been determined, there may be studies showing how long the constituents remain in blood. While a half-life analysis is particularly useful, especially if it is conducted in humans, it is not essential if there is some indication as to how long the compounds are detectable in blood. If the period of time is short, such as a few minutes to only an hour or so, it is unlikely the extract will be very effective if administered orally, unless its activity resides in an as yet unidentified metabolite or other plant constituent.
5. **Metabolism.** In general, compounds that are not metabolized are usually hydrophilic and therefore less likely to penetrate into the brain. More lipophilic components are more likely to be metabolized extensively and to cross the blood-brain barrier, making them better candidates as centrally active compounds. They are also more likely to generate pharmacologically active metabolites than hydrophilic agents. Questions should be asked about which enzymes are responsible for the metabolism of the constituents. Such information is important in attempting to determine whether the plant product might interfere with the metabolism of prescription drugs, which could result in unexpected side effects or toxicities, or a decline in the efficacy of a medication.
6. **Biological response.** Besides the knowledge about the historical use of the plant product, there may be published information on the biological response to the extract following administration to laboratory animals. An effort should be made

to determine whether the duration of any reported response matches with the time in which the constituents are present in blood. If the response lasts for several hours, but the blood levels are detectable for a much shorter period of time, then the constituents measured in blood are unlikely to be responsible for the biological action. In the absence of data from carefully controlled clinical trials, it is worthwhile to review animal studies demonstrating the biological effect of the extract. If a product is being considered because it is rumored to be a treatment for insomnia for example, such an action should be suspect if the only published animal results relate to its use for treating some other condition.

7. **Product purity and uniformity.** Because the constituents of a plant extract can vary widely among batches, even if produced by the same manufacturer, and there have been instances of product contamination, care should be exercised in selecting a supplier. When possible, domestic producers should be preferred as their manufacturing practices are more likely to be monitored by federal regulators than foreign companies. In addition, it is best to purchase the product from established retailers as they are typically more open about the location and reliability of the manufacturer, and to accepting returns if the product proves inactive or problems are encountered with its use.

Although it is unlikely that all these data will be available for most herbal supplements, and there may be times when a particular product can be obtained only from a foreign supplier, a review of the supplement in the context of this checklist will lead to a more informed decision. There are many sources for such information. These include common online search engines, including those posting published works from the scientific literature. Information can also be solicited directly from the manufacturer. However, as for any product, manufacturer claims, or the results of scientific studies underwritten by the supplier, may be subject to bias. What is best are data accumulated and reported by independent scientists having no financial interest in the outcome of the work. Of particular value are the results of experiments, both laboratory animal and human, sponsored by the federal government. There have been several such federally funded studies aimed at examining the basic and clinical pharmacological properties of popular dietary supplements.

Reference is made in succeeding portions of this work to the terms and concepts defined in this chapter. Selected dietary supplements will be discussed in detail in the context of the items listed on this checklist. Particular emphasis is placed on reviewing the published literature pertaining to the pharmacology of these products, with special attention to those studies conducted by independent investigators. By providing examples of how a pharmacologist assesses such information in drawing conclusions about safety and efficacy, the chapters on selected herbs reinforce these concepts and demonstrate the value of thinking like a pharmacologist when considering the purchase of such products.

# 4

## The Brain as a Drug Target

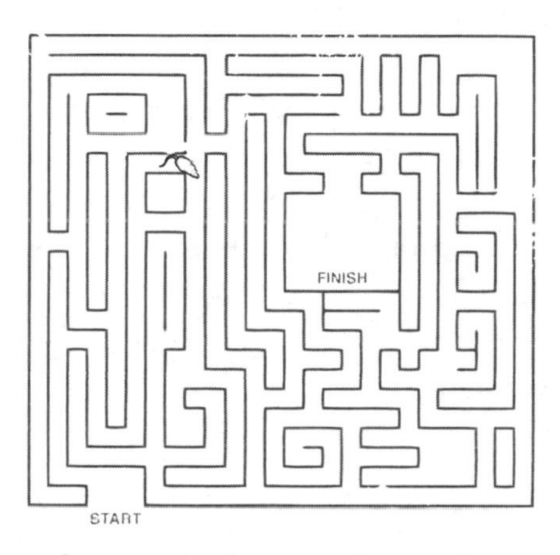


If the body is an orchestra, the brain is the conductor. All human feelings and actions are coordinated and regulated by the central nervous system, which is the composite term for the brain and spinal cord. Sensory systems, such as vision, hearing, smell, and touch, are constantly transmitting information to the central nervous system where it is almost instantaneously sorted and interpreted. The brain then sends messages to appropriate muscles to accomplish tasks as mundane as walking from the car to the office, to as complex as playing a violin or performing surgery. This ongoing monitoring of the environment continues even as the brain is being asked to perform other tasks. These include maintaining vigilance in the face of real or perceived threats, from the possibility of encountering traffic congestion to a life threatening disaster. Besides regulating the emotional behavior associated with dangers, or the anticipation of a pleasurable experience, the brain regulates the functioning of various organs and glands to correspond with these circumstances. In addition, the brain controls vital functions that require no conscious thought, such as breathing and blood pressure, and it is the organ responsible for consciousness, creativity, thought, and planning. Given its essential role in maintaining the quality of life, it is not surprising that alterations in brain function are readily perceived by the individual. This explains why early humans easily recognized plants that can affect nervous system activity. Included are those that increase energy or cause euphoria, such as central nervous system stimulants; modify perception, such as hallucinogens; cause seizures, such as convulsants; or decrease anxiety or induce sleep, such as sedatives and hypnotics. For this reason, the central nervous system effects of some plants have been reported for thousands of years. Contemporary scientific methods have been used to identify the active constituents in some of these plants, with research performed to define more precisely their actions and possible clinical benefits. In many instances, however, claims made for plant products relating to central nervous system function have not been verified scientifically, nor has an active constituent been positively identified in the extract.

Central nervous system disorders can be broadly classified into those displaying obvious and profound clinical symptoms and those involving subtle changes in emotion or affect that are often more apparent to the patient than to those observing or interacting with the individual. Examples of the former are Alzheimer's and Parkinson's diseases, amyotrophic lateral sclerosis, and schizophrenia. Depression, generalized anxiety disorder, and insomnia are conditions that fall into the second category. Although there is a great deal of overlap between these two classes with, for example, a schizophrenic also having symptoms of anxiety and depression, these conditions are considered separate entities. Because of this, drugs have been developed to lessen the symptoms of each, independent of the others.

Herbal dietary supplements are most often recommended for treating the central nervous system conditions associated with mood or affect rather than for those that involve profound changes in movement and/or cognitive ability. The reason for this is obvious. A beneficial effect in reversing paralysis or memory loss would be readily apparent to the patient and caregivers. There would be little dispute about the effectiveness of such a treatment, even in the absence of extensive clinical trials. In contrast, because depression and anxiety are sensitive to improvement by the power of suggestion, and often resolve spontaneously on their own, reports by individuals outside of well controlled clinical trials are unreliable for proving the general effectiveness of any treatment. For this reason, the benefits of a product for these conditions are unproven if they have not been rigorously and objectively demonstrated in a clinical setting. This lack of objectivity in establishing effectiveness is to the advantage of the manufacturer as some consumers will continue to hold out hope for efficacy as long as there is no proof that the product is inactive and it is generally recognized as safe. Such thinking is misguided, however, as science cannot prove a negative. Objective scientists are agnostic with regard to any hypothesis in the absence of positive proof. The consumption of a product that has not been shown conclusively to be effective for the condition of interest is based on hope, not science.

Neuropharmacology is the study of drugs that directly affect nervous system function. Neuropharmacologists perform experiments in laboratory animals and humans to ascertain whether an isolated chemical substance or plant extract modifies, for example, sensory input, organ system function, mood, behavior, or cognitive function. If positive results are obtained, the neuropharmacologist undertakes studies to determine the mechanism of action of the test substance at the cellular and biochemical levels. These data are crucial for understanding how the agent influences nerve function and for designing chemically related substances that may be more selective in terms of their site of action and, therefore, safer and more efficacious. For a consumer to assess fully the potential of an herbal supplement to beneficially affect nervous system function, it is helpful to know how drug effects on nervous system function are studied and the limitations of such work. It is also important to have some understanding of the components and organization of the central nervous system. Only then is it possible to appreciate fully how compounds can modify its functions for therapeutic gain, and the value of the brain as a drug target. With such knowledge it becomes possible to assess the likelihood that a particular product may be of value and to appreciate the risks associated with its use. Safety is of particular importance when taking any substance for central nervous system effects given the potential for catastrophic consequences when interfering with the ability of the brain to orchestrate the symphony of life.

## The Human Brain

Arguably the most complex organ in the body, the human brain is composed of billions of cells. It weighs about three pounds and has the consistency of gelatin. Approximately 100 billion cells are neurons, the elements that transmit information to, within, and from this organ. These represent only 10% of the cells in brain, with the remainder being glia, cells that nurture and protect the neurons. Structurally, neurons are composed of a cell body, which contains the nucleus. Dendrites are cellular elements that are clustered like whiskers extending from the neuronal cell body. These receive the chemical signals from adjacent neurons. Projecting also from the cell body is the axon, the part of the neuron that transmits the nerve signal to other neurons. Generally, but not exclusively, the axon terminal lies adjacent to the dendrites of another neuron. The space between the axon terminal and the adjacent cell is referred to as the synapse or synaptic cleft (see Figure 4.1). Neurotransmitters, such as serotonin and dopamine, are stored in the axon terminal and released into the synapse when the neuron is stimulated. Transmitter release is diminished if the neuron is inhibited. After diffusing across the synapse, the transmitter attaches to its receptors on the adjacent neuron, stimulating or inhibiting the affected cell (see Figure 4.1). This process is referred to as chemical transmission. It is estimated that, on average, an individual brain neuron makes synaptic contact with up to 10,000 other neurons. In this way, information is rapidly and widely disseminated throughout the brain.

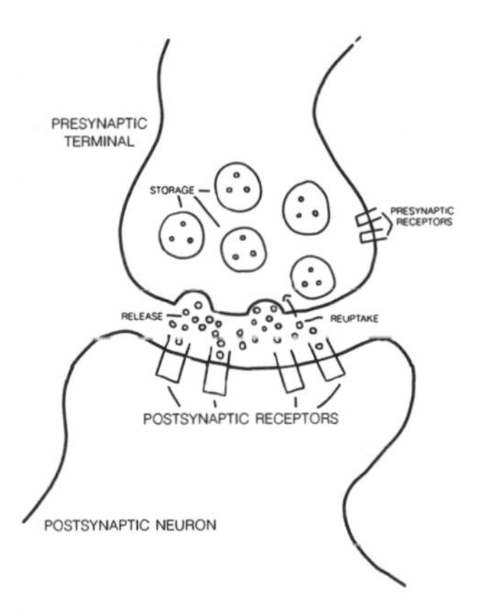


Figure 4.1 Representation of a neuronal synapse and the components of chemical transmission

Structurally, the human brain resembles those of many other species of vertebrates. Phylogenetically, the oldest areas are located in or near the brainstem, at the base of the brain. Included are the medulla, pons, and cerebellum. In isolation this unit resembles the reptile brain. A characteristic of this region, in particular the medulla, is that it controls vegetative functions, such as breathing, heart rate, and blood pressure. Destruction of this brain area is always fatal.

The next oldest area is the paleopallium (old brain). It is present in all mammals. The paleopallium encompasses structures that comprise the limbic system (see Figure 4.2).[1,2] The illustration, which represents a slice through the mid-portion of the brain, is drawn to highlight the limbic areas, not to be an anatomically accurate depiction of all regions in this section. The limbic areas include the thalamus, hypothalamus, septum, nucleus accumbens, amygdala, and hippocampus, in addition to the prefrontal, cingulate, and pyriform cortices. The prefrontal cortex, in particular, plays a major role in executive function, a distinctly human trait. Executive function relates to a human's ability to differentiate good from bad, to appreciate that actions have consequences, and to plan for the future. The limbic system, or circuit, is also responsible for emotion and memory. It is important for self preservation as it is this circuit that enables an individual to appreciate the danger associated with, for example, confronting a hungry lion, or riding with a drunk driver. Alterations in this system are thought to be responsible for some symptoms of psychiatric disorders, including major depression, anxiety, and psychosis. The memory deficit characteristic of Alzheimer's disease is associated with alterations in the hippocampus.

The neopallium is the most recently developed region of the human brain. It is composed of large areas of the cerebral cortex, and some subcortical regions. Indeed, the greater size of the cerebral cortex in humans is the major difference between the human and other mammalian brains. The neopallium is the brain region responsible for distinctly human activities, such as abstract thought, speaking, reading, and writing. While damage to this brain area is not necessarily fatal, these human functions can be compromised, as is sometimes seen after a stroke.

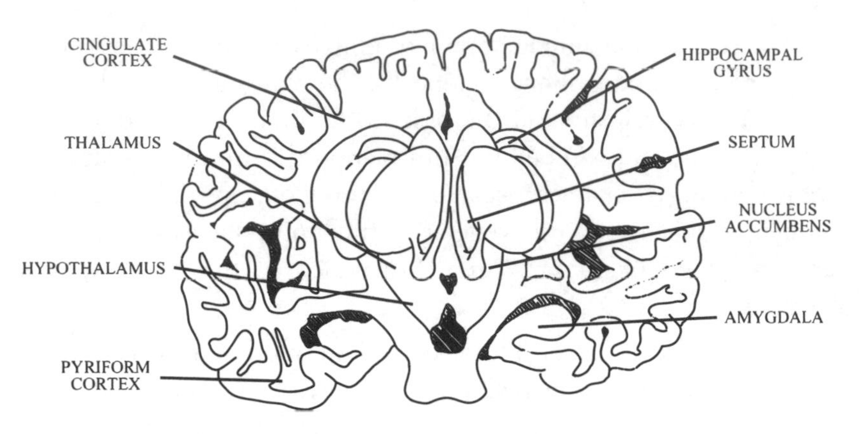


Figure 4.2 Composite diagram of a cross-section of the vertebrate brain

Anatomically and functionally the cerebral cortex can be divided into three distinct regions. The sensory cortex receives and processes information from sensory organs. Signals relating to vision, sound, taste, and touch are sent to, decoded, and interpreted in the sensory cortical regions. After processing these data, the brain activates, or inhibits, the primary motor area of the cerebral cortex, which sends neuronal projections down to the brainstem and motor pathways in the spinal cord to initiate, maintain, or inhibit skeletal muscle activity. The sensory information is also transmitted to the third cortical region, the association cortex, which is comprised of all the cortical areas not included in the sensory and motor cortices. The association areas are responsible for the conscious interpretation of the information and for making decisions as to how to react to the findings. Because of their highly developed association cortex, humans are considered to be more capable than other animals of overriding, or controlling, instinctive behaviors.

While it is tempting to associate individual brain areas with particular functions, it is now believed that behaviors result from the coordinated activity of brain circuits involving several brain regions. The limbic system is one such circuit. Another is the extrapyramidal system, which is responsible for maintaining posture, regulating motor reflexes, and coordinating movement. Brain areas associated with the extrapyramidal system include the vestibular nuclei, cerebellum, thalamus, substantia nigra, and the corpus striatum, which is comprised of the caudate nucleus, putamen, and globus pallidus. Destruction of the nerve pathway that originates in the substantia nigra and terminates in the corpus striatum is a hallmark of Parkinson's disease. This neuronal damage is responsible, in part, for the abnormal gate, postural dysfunction, and tremor that are characteristic of this disorder. Thus, interruption of a brain circuit can have significant, and permanent, clinical consequences. Drugs can also interfere with these pathways, which may have therapeutic benefit if the system is overactive, or may be responsible for side effects. For example, some classes of antipsychotics are known to cause Parkinson's disease-like symptoms because they block the substantia nigra pathway that innervates the corpus striatum.

By understanding the functional relationships among brain regions, and then identifying the brain areas associated with a particular neurological or psychiatric disorder, it is possible to devise surgical or pharmacological approaches to treat these conditions. With regard to the use of drugs or natural products, it is important to be able to target the active agent to the brain circuit and, if possible, the precise neuronal pathway, responsible for the symptoms. Given the complexity of the brain, and the interactions among brain regions, a generalized disruption of neurotransmission could be catastrophic. At the very least it could be associated with many untoward side effects. Fortunately, the chemical nature of neurotransmission provides a means for targeting individual brain pathways, thereby making it possible to modify selectively central nervous system activity.

## Chemical Neurotransmission

The characterization of chemical neurotransmission advanced considerably the understanding of how drugs and other chemicals affect brain function. In addition, defining how neurons communicate with one another made it possible to identify new targets for developing drugs for treating neurological and psychiatric disorders. Because of this, knowledge of the fundamentals of chemical transmission is useful for appreciating how drugs and dietary supplements may modify brain function and for understanding their proposed mechanisms of action.

Transmitters are chemicals synthesized in neurons. Two dozen or so chemical substances have been identified as brain neurotransmitters. These agents are the primary means of communication between brain cells. Transmitter vocabulary is limited, however, in that a given agent either enhances or decreases the firing of an adjacent neuron. By affecting the excitability of the neuron, transmitters influence the way the receptive cell modifies the activity of neighboring neurons. In most cases the transmitter is stored in packets, called vesicles, in the nerve terminal (refer to Figure 4.1). When the neuron is stimulated an impulse travels along the axon causing the vesicles to fuse with the presynaptic axon terminal membrane, to open, and then to discharge the stored transmitter into the synapse (refer to Figure 4.1). Once released, the transmitter diffuses across the synaptic cleft and attaches to its receptors on the adjacent postsynaptic neuron. Typically, the postsynaptic membrane is located on the dendrites of an adjacent cell. Attachment of the transmitter to its receptor enhances, in the case of excitatory neurotransmitters, or inhibits, in the case of inhibitory neurotransmitters, the activity of the postsynaptic neuron. If the neuron is stimulated, the impulse generated facilitates the release of the neurotransmitter from its axon terminals to transmit a signal, which is either excitatory or inhibitory, to adjacent neurons. By slowing the firing of the receptive neuron, inhibitory transmitters decrease transmitter release from the adjacent cell which, in turn, influences the firing of neighboring neurons. Thus, overall brain activity represents a delicate balance between excitatory and inhibitory transmitters.

Besides being localized on postsynaptic membranes, neurotransmitter receptors may be found on presynaptic terminals (refer to Figure 4.1). When activated these presynaptic receptors slow the release of a transmitter from that terminal by causing intracellular biochemical changes that reduce vesicular attachment to the membrane. Thus, if an excessive amount of neurotransmitter is being liberated from a presynaptic nerve terminal, some will leak out of the synapse and stimulate the presynaptic receptors, slowing further transmitter release in an attempt to return the system to equilibrium. It is also possible that an axon from a neighboring neuron may synapse with a presynaptic, rather than postsynaptic, membrane. This is referred to as an axo-axonic, as opposed to axo-dendritic, synapse. If there is an axo-axonic interaction the rate of transmitter release from a terminal is regulated not only by the firing rate of that neuron, but also by the activity of an adjacent cell releasing its transmitter onto the presynaptic receptors.

Once a neurotransmitter activates its receptor it is released back into the synapse. Some transmitters are destroyed by metabolizing enzymes located at or near the synapse. Many are also transported back (reuptake) into the presynaptic terminal where they may be metabolized intracellularly or re-stored in vesicles for future use (refer to Figure 4.1).

The elements involved in synaptic transmission provide many targets for manipulating this process pharmacologically. For example, brain neurotransmitter function can be modified by agents that inhibit the synthesis, metabolism, or storage of neurotransmitters, or their reuptake into the presynaptic terminal. Agents are also known that enhance the release of certain neurotransmitters. Commonly drugs, and possibly natural products, directly interact with neurotransmitter receptors, either activating (agonists) or blocking (antagonists) these sites. Brain neurotransmission is also affected by drugs that interact with neuronal ion channels or ion transporters. The passage of ions across the neuronal membrane is essential for transmitting nerve impulses and for maintaining a healthy intracellular environment. Activation of some receptors can affect ion channel activity or ion transport, thereby modifying cellular excitability and stability. Likewise, chemical agents can modify brain function by attaching directly to these sites.

There are many other potential drug targets in neuronal tissue besides those involved directly with chemical neurotransmission. Included are enzymes, such as kinases and phosphatases, responsible for modifying protein function, and intracellular receptor systems, such as those for certain hormones, which regulate gene expression in the cell nucleus.

## Neurotransmitter Systems

Most transmitters interact with a particular family of receptors. For example, by definition, the neurotransmitter dopamine only activates dopamine receptors, yet there are molecular differences among these sites that make it possible to design drugs that activate, or block, only certain types of dopamine receptors. Moreover, molecularly distinct receptors for a given neurotransmitter are differentially localized throughout the brain. The existence of such neurotransmitter receptor subtypes makes it possible to design drugs that will selectively affect a particular neurotransmitter system in a targeted brain region.

Examples of brain neurotransmitter systems known to be influenced by drugs and, in some cases, natural products, are shown in Table 4.1. Among these are γ-aminobutyric acid (GABA) and glutamic acid transmitter pathways. Quantitatively at least, GABA and glutamic acid are believed to be the two most important transmitters in the brain. It is estimated that up to 40% of the neurons in the central nervous system release GABA, with another 40% utilizing glutamic acid. Thus, 80% of the neurons in the brain and spinal cord are either GABAergic or glutamatergic. The other 20% of the neurons utilize one, or more, of the other neurotransmitters. As GABA is an inhibitory neurotransmitter, and glutamic acid an excitatory agent, and because both are found in neurons distributed throughout the central nervous system, it is the balance between these two systems that maintains global brain function. By contrast, neurons utilizing dopamine or serotonin as their transmitter are more highly localized in the brain, in particular in limbic regions, and, therefore, unlike GABA and glutamic acid, have more circumscribed, discrete effects on brain function. Given these facts it is likely that most, if not all, central nervous systems disorders involve, either directly or indirectly, alterations in GABA or glutamic acid transmission.

**Table 4.1** Some Brain Neurotransmitter Targets for Drugs and Natural Products

| Neurotransmitter | Clinical Condition* | Drug/Natural Product |
|---|---|---|
| Acetylcholine | Cognitive and memory dysfunction, Alzheimer's and Parkinson's diseases | Benztropine, donepezil/ **nicotine, epipatidine** |
| Dopamine | Depression, schizophrenia, addiction | Haloperidol, amphetamine/ **cocaine** |
| Serotonin | Anxiety, depression, obsessive compulsive disorder | Fluoxetine/**psilocybin** |
| Glutamic Acid | Pain, anesthesia | Ketamine/**kainic acid** |
| GABA | Anxiety, seizures, insomnia | Diazepam, phenobarbital/ **bicuculline** |
| Adenosine | Somnolence | **caffeine** |

*Relates to some of the clinical uses of agents known to interact with the designated neurotransmitter system.

The benzodiazepines, such as diazepam (Valium®), increase the sensitivity of a particular class of GABA receptors to the action of GABA, thereby enhancing transmission at GABA synapses. Likewise, barbituates, such as phenobarbital, directly interact with some GABA receptors to prolong the response when the site is activated by the neurotransmitter. Because GABA is an inhibitory neurotransmitter, by enhancing GABAergic transmission these drugs decrease central nervous system activity, making them useful for relieving anxiety and for inducing sleep and anesthesia. High doses of such agents can be fatal if they cause inhibition of brainstem cells responsible for regulating the cardiovascular and respiratory systems. In contrast, as would be expected, the GABA receptor antagonist bicuculline, an alkaloid present in extracts of Dicentra cucullaria, known commonly as Dutchman's britches, and Adlumia fungosa, a member of the poppy family Papaveraceae, is a convulsant.

As for the glutamic acid system, ketamine is a drug that inhibits a certain type of glutamic acid receptor. In so doing, it causes amnesia and analgesia, making it a useful fixed anesthetic for certain types of patients. As ketamine also causes hallucinations, its use is limited. Kainic acid, a natural product obtained from a Japanese seaweed (Digenea simplex), stimulates certain members of the glutamic acid receptor family. Because glutamic acid is an excitatory neurotransmitter, it is not surprising that kainic acid causes seizures. It is also neurotoxic, with exposure causing nerve cell death.

A great deal of effort has been expended on developing drugs that selectively modify the GABA and glutamic acid transmitter systems. Given the importance of these two neurotransmitters in brain function, attempts are often made to explain a purported central nervous system action of a herbal supplement on an interaction with one or both of them.

Adenosine, which is released from both neurons and glia, is an inhibitory neurotransmitter in the brain that can also modulate the action of other transmitters.[3] Activation of the adenosine system decreases central nervous system activity and causes drowsiness. Caffeine, a natural product found in many beverages, such as coffee and tea, is an adenosine receptor antagonist. This explains the stimulant actions of caffeine, and its popularity as an aid for maintaining wakefulness.

Serotonin is a neurotransmitter localized to neurons that originate in the brainstem but that make synaptic contact in higher brain regions, in particular those associated with the limbic system.[4] There are also serotonin synapses in the spinal cord. Drugs, such as fluoxetine (Prozac®), inhibit the presynaptic reuptake of serotonin, thereby prolonging the action of this neurotransmitter. Such agents are used to treat clinical depression and obsessive compulsive disorders. Psilocybin is a serotonin receptor agonist. An hallucinogen, psilocybin is present in various species of mushroom (genus Psilocybe). Its ritualistic and recreational uses date back thousands of years in southern European and Central American cultures.

Acetylcholine is a neurotransmitter that is widely distributed throughout the brain. Inhibition of certain types of acetylcholine receptors causes hallucinations and memory impairment, whereas activation enhances memory and cognition. Brain acetylcholine receptors are targets for the development of analgesics, antipsychotics, and antidepressants. It is known that the memory dysfunction associated with Alzheimer's disease is due, at least in part, to the destruction of an acetylcholine pathway in the hippocampus. Accordingly, drugs that inhibit the metabolism of acetylcholine, such as donepezil, are employed for the treatment of this condition. As the neuronal loss responsible for the symptoms of Parkinson's disease results in an overabundance of cholinergic activity in the brain, cholinergic receptor antagonists, such as benztropine, are sometimes prescribed for this condition. Several natural products, including nicotine and epibatidine, are known to stimulate directly certain types of acetylcholine receptors in the central nervous system. Nicotine is an alkaloid present in plants of the genus Solanaceae, which includes tobacco. By stimulating brain acetylcholine receptors, nicotine increases alertness, and can cause nausea and vomiting, especially in the uninitiated. Epibatidine is a compound extracted from a South American frog (Epipedobates tricolor). Epibatidine induces a profound analgesia by stimulating directly a subclass of acetylcholine receptors in brain. Its toxicity precludes its use by humans.

Dopamine is a transmitter best known for its involvement in the symptoms of Parkinson's disease and schizophrenia. Localized in a discrete set of neurons that project to limbic, endocrine, and motor areas of the brain, alterations in dopaminergic transmission have significant effects on a variety of behaviors and on motor function. Antipsychotic drugs, such as haloperidol, block a certain class of dopamine receptors, dampening the symptoms of schizophrenia. This suggests that these symptoms are due to an overactive dopamine system in certain brain regions. Amphetamine, a central nervous system stimulant, works in part by enhancing the release of dopamine in the brain. It can exacerbate the symptoms of schizophrenia. Amphetamine also facilitates the release of other neurotransmitters, including serotonin and norepinephrine. Cocaine, which is contained in the leaves of the cocoa plant Erythroxylon cocoa, has been used for centuries for its central nervous system stimulant properties. Cocaine inhibits the reuptake of dopamine and other brain neurotransmitters.

This list represents only a few of the various brain transmitters that can be modified for therapeutic gain, and only a small fraction of the drugs and natural products known, or believed, to affect these systems. Nonetheless, it illustrates that chemically induced modification in brain transmitter function can cause profound changes in behavior, memory, cognition, and alertness. These data also suggest that drugs and natural products can have selective effects on individual neurotransmitter systems, making it possible for them to target certain brain circuits. Information such as this has led many to examine the effects of herbal supplements on brain transmitters. The identification of such an effect suggests that components of these products can influence central nervous system function, and can predict the type of behavioral or therapeutic response that might be anticipated when consuming these products.

## Behavioral Assays

Behavioral responses to new chemical compounds are often first identified and characterized in laboratory animal tests. Animal behavioral assays are also utilized to characterize the central nervous system effects of herbal supplements thought to influence central nervous system function. In some experiments, drug candidates or plant extracts are given to normal animals, generally rats or mice, to determine whether they modify animal behavior. For example, when administered at the proper dose to rats, amphetamine has a significant effect on motor activity and induces stereotyped, or obsessive, behaviors. In contrast, haloperidol, a dopamine receptor antagonist, causes catalepsy, a type of muscle rigidity, and blocks the response to amphetamine. Such behavioral and pharmacological effects are characteristic responses with activating or inhibiting certain dopamine systems in the brain.

Regardless of whether an effect is observed on normal behavior, tests may be conducted to determine whether a newly synthesized chemical substance or a plant product can alter animal behavior in response to certain challenges. The results of this type of analysis can suggest how these agents might affect human behavior, or what type of neurological or psychiatric condition might benefit from the use of this agent. Given the differences between human brains and those of laboratory animals, the results of such tests only suggest, but do not prove, a possible clinical response. For example, compounds that cause seizures in laboratory animals will usually be convulsants in humans as well. On the other hand, agents that may appear to be analgesics in animal tests do not always display such activity in humans. Effects in animals are even more difficult to translate to humans when studying behaviors used to assess emotion, affect, cognition, and memory. Nonetheless, such tests are useful in helping to determine whether a drug candidate, or plant product, penetrates into the brain, its duration of action, and its potency in inducing some type of behavioral response. A test agent that fails to alter animal behavior at reasonable doses is unlikely to have any effect on central nervous system activity in humans. Provided below are some examples of laboratory animal tests employed to study the central nervous system effects of drug candidates and herbal supplements.

A number of animal tests are used to determine whether a chemical will reduce (anxiolytic) or cause (anxiogenic) anxiety. One of the more common assays is the elevated plus maze.[5,6] The test apparatus consists of an enclosed box in the shape of a plus sign with four corridors radiating from the central area. These corridors are roughly the size of a rodent burrow. The ends of two of the corridors abut a solid wall, whereas the ends of the other two are open. As the entire apparatus is elevated about two feet from the floor, the rat or mouse will not jump from the open ends. While rodents are prone to explore new territory, they are apprehensive about being exposed, especially on an elevated platform. Therefore, a rat or mouse placed into the central portion of the maze will generally spend more time wandering in the closed than in the open corridors. If animals administered a test substance spend more time than normal in the open than in the closed corridors it is concluded that the compound or plant extract reduces anxiety. Conversely, a compound or plant mixture that causes animals to spend more time than normal in the closed than in the open corridors is thought to be an agent that increases anxiety. As with all animal tests, however, other explanations for observed behavior must be considered. For example, a test agent that compromises vision might be interpreted as being an anxiolytic because the animal spends more time than normal in the open corridor. However, the animal may behave this way simply because he does not realize it is open at the other end. Also, a compound that causes muscle weakness, but that has no effect on anxiety, might be interpreted as being an anxiolytic or anxiogenic agent if the animal remains for an abnormally prolonged period in either an open or closed corridor, respectively, because of difficulty in moving from one to the other. Accordingly, these types of assays must be designed to control for effects having nothing to do with anxiety that may influence the results and their interpretation.

There is interest in developing new animal models for studying the potential value of chemical agents that modify behavior. As more information is gathered on the genetic basis of central nervous system disorders, animal models, usually involving mice, can be generated by inducing similar alterations in laboratory subjects. An example is the development of a mouse model of autistic behavior that results from a specific genetic manipulation of these animals.[7]

A rat model of Alzheimer's disease has been developed that involves an intracerebral injection of beta amyloid, a protein present in excessive quantities in the brains of Alzheimer's patients. Following injection of beta amyloid into the rat brain, the animals display some of the symptoms of this condition, including deficiencies in learning and memory.[8] Several tests of cognition are used to evaluate learning and memory in laboratory animals. In the Morris water maze assay, for example, a rodent is placed into a water-filled pool that contains a platform that the animal can use as a refuge.[9] When tested over several days, the animal becomes more proficient at locating the platform so as to escape the water more quickly. Learning is measured by noting the speed with which the animal learns to escape. Memory is measured by studying the retention of this skill over time. Test agents, such as drug candidates or plant extracts, that reduce the number of trials needed to locate the escape platform are thought to facilitate learning, whereas those that lengthen the amount of time the skill is retained are believed to enhance memory.

A simpler test is reversal of scopolamine-induced memory impairment.[10] Scopolamine, an acetylcholine receptor antagonist, is a natural product found in plants from Solanaceae, such as jimson weed and henbane. Scopolamine is known to block short-term memory in humans and other animals. A test substance capable of blocking or reversing scopolamine-induced amnesia in rodents may potentially enhance memory in humans.

A number of publications describe the various laboratory animal behavioral tests used to examine drug candidates.[11-13] In all cases, the end point is a quantitative assessment of some behavioral task, which may or may not bear any obvious relationship to a human condition. Often, the response to the test compound is compared to results obtained with a drug known to be effective in treating the disorder of interest. However, as illustrated previously, few, if any, animal models of behavioral disorders fully recapitulate the human condition. This is not surprising given the differences between human and rodent brains, especially with regard to cerebral cortical function. Also, as the cause and the underlying pathology of many neurological and psychiatric disorders is unknown, it is impossible to replicate the disorder in laboratory subjects. Given these difficulties, results from laboratory animal behavioral tests are not definitive in terms of predicting responses in humans. Prudent investigators employ multiple animal tests for, say, anxiety or depression, before making any conclusions about the likelihood of the drug candidate or plant product having any beneficial effect in humans. Drawing conclusions about the utility of an herbal supplement for treating a central nervous system disorder is therefore complicated by the fact that few undergo rigorous, multiple animal tests for demonstrating efficacy as a treatment for a central nervous system disorder.

## Clinical Studies

The most conclusive data demonstrating effectiveness in humans are derived from well-designed clinical trials. There are several different types of clinical observations. The least reliable are anecdotal reports of an unexpected response in a few patients. While such findings can provide important leads, they must be confirmed in large scale clinical studies designed specifically to test the hypothesis generated by the anecdotal finding.

Data are also derived from open-label clinical trials. While the number of subjects may be greater than is the case for an anecdotal report, and the study is designed to test a hypothesis, open-label means that both the investigator and the patients know who is receiving the test agent. This can lead to a bias in the participant reports. While the results of appropriately conducted open-label studies should not be ignored, they must be interpreted with caution and considered preliminary.

The most definitive clinical data are the results of large scale trials, which might involve hundreds to thousands of subjects, which are multicentered, meaning they are conducted simultaneously at several institutions around the country or the world. To be definitive, such trials should be, when possible, double-blind, crossover studies. Double-blind indicates that neither the clinician nor the patient knows who is receiving the test agent, a known drug, or a placebo. In a crossover study some of the patient volunteers receive the test substance for a period of time, while others do not. After a predetermined period of time, the groups are switched, or crossed over, in terms of what they are administered. This makes it possible to determine whether any effect observed with the test substance dissipates over time, which would be expected, and whether a positive clinical response in the two groups is associated with the administration of the test substance. The fact that the study is double-blind eliminates the potential for the bias associated with open-label trials, and the large number of subjects, and multiple clinical sites, maximizes the likelihood that any positive finding will be clinically and statistically meaningful, rather than a random event. The results of such studies, which are very expensive and time consuming, are considered the most definitive in assessing the response to a drug candidate or natural product. While all FDA approved drugs must undergo such testing, herbal supplements generally do not. Because there have been natural product supplements used for clinical purposes, some have undergone this type of testing. When such data exist they should be carefully considered when deciding whether a particular natural product may be of value.

Clinical studies aimed at assessing the effects of a test substance, or natural product, on emotional behavior are particularly difficult to design and evaluate. A major reason for this is the lack of objective biological markers for these conditions. For example, when testing a new antihypertensive drug candidate, antidiabetic agent, or antibiotic, clinical effectiveness is monitored by measuring signs such as blood pressure, blood sugar, or bacterial counts, respectively. There are no laboratory tests for monitoring psychiatric disorders, such as major depression, schizophrenia, generalized anxiety, and insomnia. Instead, these conditions are diagnosed mainly on the basis of symptoms, which can be difficult to quantify consistently over time. The assessment of drug efficacy in the treatment of some psychiatric or behavioral disorders is complicated further by the fact that some symptoms, or the conditions themselves, resolve on their own. Given the lack of objective laboratory tests to monitor these disorders, symptoms are documented and monitored regularly by the patient and caregivers to assess the effectiveness of a potential medication during a drug trial. A number of symptom scales have been developed for this purpose. Examples include the Symptom Check List-90, the Hamilton Anxiety Scale and the Hamilton Depression Scale.[14-16] Cognitive dysfunction, such as that associated with Alzheimer's disease, is quantified using the Neuropsychiatric Inventory and Alzheimer's Disease Assessment Scale. By cataloguing symptoms and subjectively ranking their severity, it is possible to generate numerical values that can be used to determine whether a particular treatment is of benefit. While such assessments are less reliable than objective laboratory tests, they have proven useful in the development of psychotherapeutics, and in examining the effectiveness of products, such as herbal supplements, that are thought to be of value in treating these conditions.

Given the difficulties associated with determining the effect of test substances on emotional behavior, it is critical that, when possible, the clinical response be assessed in comparison with patients receiving only a placebo, or sham treatment. In drug trials, the placebo is a tablet, pill, or syrup that appears to the eye and palate to be identical to the preparation containing the agent under investigation. However, unlike the test substance, the placebo contains only inert material. It would be expected, therefore, that those subjects who are unknowingly receiving the placebo will not display any improvement in their condition. By comparing the response in patients administered the test agent to those receiving the placebo it is possible to determine the statistical probability that the experimental treatment displays efficacy as a treatment for the condition. Inclusion of a placebo group in a clinical trial is complicated by the fact that these patients are being denied treatment. For conditions that can be fatal, such as with severe depression when suicide is a risk, withholding treatment is unethical. In these situations, the comparison group may be administered an established antidepressant to see whether the test agent is as effective as the conventional drug.

Another challenge with placebos is the fact that patients often report a positive response when taking them. The placebo response rate can be especially high when treating emotional disorders. It is estimated that, depending on the condition being studied, up to 30% of those receiving a placebo may report benefitting from taking the inert material. This complicates the testing of psychotherapeutics in particular since, in many cases, the clinical response to these drugs is observed in only 50% to 60% of the subjects in the study. If symptoms in 30% of a comparison placebo group improve, and it is known that a certain fraction of the patient population would have improved during the time of the study without any medication at all, the actual clinical benefit of the treatment under investigation may be difficult to demonstrate. This is true for drugs as well as for herbal supplements. For these reasons, many patients must be studied in well-designed clinical trials conducted by numerous, independent, experienced investigators to prove that a therapy qualifies as a treatment for certain central nervous system disorders. Most herbal supplements have not yet undergone such rigorous testing. For this reason, many pharmacologists and others remain skeptical about rumors concerning the effectiveness of these products as treatment for central nervous system disorders.

# 5

# Ginkgo (Ginkgo biloba)

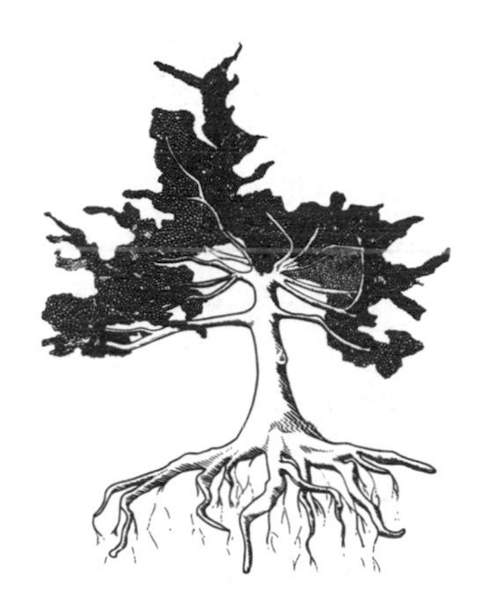

One of the most popular herbal supplements, extracts of the ginkgo leaf are recommended for the treatment of a variety of conditions. In particular, they are thought to be of benefit for enhancing memory and cognition. The popularity of this product is demonstrated by the fact that annual sales of ginkgo exceed $1 billion worldwide, with more than $100 million of this spent in the United States alone.

Ginkgo has been widely available in the Western hemisphere for the past 40 years, first in Europe and then in the United States. The relatively recent interest in this leaf extract is notable because ginkgo, also known as maidenhair or kew, is one of the oldest living species of tree. Fossil evidence suggests that ginkgos were present throughout the world up to 200 million years ago, but were rendered nearly extinct during the last ice age, surviving only in Asia. Reintroduced to the West in the sixteenth century, modern ginkgo is the sole survivor of the Ginkgophyla division of the Ginkgoaceae family. A dioecious species, the female ginkgo produces a plum-like fruit that emits an offensive odor due to the presence of butanoic and hexanoic acids.

The male ginkgo is preferred for ornamental planting in part because of the unpleasant odor of the rotting fruit.

Although the ginkgo was present throughout human evolution, the possible medicinal value of its fruit, seeds, and leaves was not recorded until some 800 years ago. One of the earliest extant publications on its therapeutic potential is Lan Mao's *Dian Nan Ben Cao* where it is suggested that the leaves be used as a topical treatment for freckles, head sores, chilblains, and wounds. The first known mention of systemic use, for the treatment of diarrhea, appeared in the fourteenth century with the publication of Liu Wen-Tai's *Ben Cao Pin Hui Jing Yao*. At about the same time, Li Shih-Chen in *Pen Ts'ao Kang Mu* proposed ginkgo seeds as a remedy for a host of conditions, including cough, asthma, and worm infections. It was not until the 1960s, however, that ginkgo leaf extract was introduced as an herbal remedy in Western Europe. Its popularity in the United States dates from the 1980s.

It is possible the delay in documenting the possible medicinal value of ginkgo was because of its limited geographical distribution and the unpleasant aroma of its fruit. The former seems unlikely as the Chinese were publishing descriptions of herbal remedies, such as in the *Shen Nong Ben Cao Jing*, as early as 2800 BC. As for the possibility that the foul odor lessened enthusiasm for consumption of the fruit, seeds, and leaves, there is written evidence suggesting ginkgo seeds were a food source in China from at least 200 BC. This indicates these plant products have been ingested for more than 2000 years. It seems more likely that the ancients were slow to appreciate the possible medical value of the ginkgo because responses to the application or consumption of its seeds, leaves, fruit, or their extracts, are subtle and unpredictable. This differentiates the ginkgo from other natural products, such as opium, which display a rapid, dramatic, and consistent effect on central nervous system function. Even with the most modern tools available for assessing cognition and memory, there is still debate about whether gingko extract improves brain function. A careful analysis of the literature on the chemical composition of gingko extract, and on what is known of its pharmacokinetics and pharmacodynamics, provides some insights into this ongoing controversy.

## Botany

Today, leaf extracts of ginkgo are the most commonly used herbal remedies, although the seeds are consumed for this purpose in some countries. Following the last ice age, the ginkgo and related species continued to grow wild in what is now China. With the arrival of humans, ginkgo survived under cultivation while all other species of Ginkgoaceae became extinct. The Chinese ginkgo was subsequently exported to other countries as an ornamental tree, reestablishing its presence throughout the world.

The name ginkgo comes from the Japanese term ginkyo, meaning silver apricot, an apt description of the fruit produced by the female of the species. Biloba refers to the shape of the leaf, which resembles a small fan having two lobes. The tree can grow to over 100 feet, with individual specimens living for 1,000 years or more. The seeds are contained in cherry-like seed heads. Botanically, ginkgo is a Gymnosperm. Like pines and other members of this group, its mature seeds are not enclosed in an ovary. Yews are considered the most closely related living relative of the gingko. A very hardy plant with a significant resistance to disease, ginkgos are found in temperate regions throughout the world. The appearance in the West of the cultured variety was recorded in Europe in the seventeenth and in North America the nineteenth centuries.[1]

## Therapeutic Uses

Given its long absence from Europe, the ginkgo is not included in Western writings up through the Renaissance. In Chinese medicine, both the seeds and the dried leaves were used to treat numerous conditions. In the *Pen Ts'ao*, Li Shih-Chen recommends the ripe seed be taken orally to reduce cough and dyspnea, and for the treatment of asthmatic bronchitis.[2] In addition, these seeds were used in traditional Chinese medicine for managing leukorrhea, a vaginal condition, and enuresis, or bedwetting. Recorded side effects and toxicities for the seeds include muscle spasm, seizures, skin irritation, and kidney inflammation.

The use of ginkgo leaves and their extracts for therapeutic purposes is a more recent development. A modern compilation of

Chinese herbals includes ingestion of the ginkgo leaf or extract for the treatment of Parkinson's disease, migraine, atherosclerosis, hypercholesterolemia, and chronic bronchitis.[3]

In the West, leaf extracts are taken primarily to improve memory and cognition, especially in the elderly. Several of the more popular recommended uses as a treatment for central nervous system disorders include memory loss in general, the memory deficits associated with Alzheimer's disease, and a condition referred to as mental fatigue (see Table 5.1). Ginkgo is also reported to improve the mental health of those afflicted with multiple sclerosis, and to have some positive benefit in the treatment of glaucoma, macular degeneration, and tinnitus, or ringing in the ears. Other advertised indications are allergic inflammation, asthma, hardening of the arteries, Reynaud's disease, psoriasis, vitiligo, male infertility, and generalized aging (see Table 5.1). While this list is not exhaustive, it illustrates the wide range of purported beneficial effects of this extract.

**Table 5.1** Some Recommended Indications for Ginkgo Biloba Extract

| | |
|---|---|
| Aging | Male Infertility |
| Allergic Inflammation | Memory Loss |
| Alzheimer's Disease | Mental Fatigue |
| Asthma | Multiple Sclerosis |
| Glaucoma | Psoriasis |
| Hardening of the Arteries | Reynaud's Disease |
| Intermittent Claudication | Tinnitus |
| Macular Degeneration | Vitiligo |

It is difficult for a pharmacologist to assess the effectiveness of ginkgo as a treatment for conditions such as mental fatigue for which the pathology is unknown and the symptoms purely subjective. It is also not possible to assess effects on aging without identifying the specific aspect of the aging process to be studied. In contrast, the efficacy of ginkgo extract as a treatment for defined clinical conditions, such as Alzheimer's or Reynaud's diseases, and as a means for lessening symptoms, such as memory loss, can be examined critically.

It is notable that the earliest written reports on the therapeutic uses of ginkgo do not generally include mention of effects on central nervous system function. Rather, emphasis was placed on its possible value as a treatment for respiratory conditions, such as asthma, and vascular disorders, such as intermittent claudication, which is hardening of the arteries in the legs. This suggests that early ginkgo preparations displayed no obvious effects on brain function.

Interest in ginkgo leaf extract as a palliative, if not a remedy, for memory loss was stimulated in the latter half of the twentieth century by the belief that it induces vasodilation, and therefore may increase blood flow to the brain. Interest in this use was fostered by the production and sale of a ginkgo extract, EGb 761, by Dr. Willmar Schwabe GmbH & Company in Germany. Research was performed, some underwritten by the company, to examine its effects in humans and to assess its mechanism of action. While positive results were published on the clinical effectiveness of EGb 761 in enhancing memory and cognition, these conclusions were often based on data from uncontrolled trials, anecdotal reports, case studies, or small statistically underpowered studies. The popularity of ginkgo extract grew considerably, especially in the United States, when Dr. Elias James Corey mentioned his work on the total chemical synthesis of ginkgolide B, a constituent of the extract, when accepting the 1990 Nobel Prize in Chemistry. While Dr. Corey made no mention of the possible therapeutic benefits of ginkgolide B, nor endorsed its use as an herbal product, his chemical interest in the compound was taken as a validation of its clinical potential.[4] Sales of ginkgo extract increased substantially in the 1990s as it became a popular herbal supplement for enhancing memory in the elderly and in those experiencing cognitive decline, regardless of the cause.

## Constituents

A hallmark of conventional pharmaceuticals, whether prescription or over-the-counter, is that the precise amounts of all chemical components and their pharmacological properties are known. This includes not only the active component, but also any biologically inert materials included in the product as preservatives or to enhance solubility, taste, or absorption of the drug. This is not the case with preparations of gingko. There are scores, if not hundreds, of chemicals in gingko products, with the exact number and type depending on the extraction and purification procedure. Typically, commercial ginkgo products contain an extract resulting from several different purification steps to enhance the concentration of some constituents, and to lower that of others.[5] After the extraction procedure, the solvent is removed and the dried powder sold to the consumer. As these processing steps may vary among manufacturers, the chemical composition differs among producers. Indeed, as with wines, variations in constituents among batches would be anticipated even with the same extraction process because the relative quantities of the various chemicals in the leaves are affected by many factors, such as the growing conditions, the time of year the leaves are harvested, and the age of the tree.

The aim in preparing most ginkgo leaf extracts is to have a powder composed of 6% terpene trilactones and 24% flavonol glycosides, with only a trace of ginkgolic acids.[5] This mixture of terpenes and flavonols is referred to as the standardized extract. The remaining constituents of this preparation, composing roughly 70% of the total, are generally unidentified in individual preparations. This large, uncharacterized, component is known to include various classes of organic compounds, such as proanthocyanidins, carboxylic acid derivatives, polyphenols, catechins, carbohydrates, alcohols, ketones, alkylphenols, and non-flavonol glycosides. The standardized extract also contains a host of undefined, high molecular weight compounds and inorganic molecules. It is estimated that approximately 13% of the standardized powder has never been identified.[5] Besides its commercial use, the standardized extract is often employed for preclinical and clinical studies.

As the list above includes only chemical classes, the actual number of individual agents in ginkgo powder is unknown. It is believed that any therapeutic benefit derived from the consumption of the standardized extract is due to the actions of certain flavonoids and terpenes, some of which have been chemically characterized. While the ginkgo leaf flavonoids receiving the most attention are kaempferol, quercetin, and isorhamnetin, at least 40 others in the preparation are present in smaller quantities. Chief among the ginkgo terpenes are ginkgolides A, B, C, J, and M, and bilobalide. Of these, ginkgolide B has been examined most thoroughly as it is believed to be one of the most active ingredients in the preparation. Some other individual compounds identified in the ginkgo extract are shikimic, vanillic, ascorbic and p-coumaric acids, as well as sitosterol and stigmasterol.

Thus, only a small fraction of the individual chemical compounds present in the standardized ginkgo product are known. This fact, plus the variations that occur in the concentrations of these constituents among the commercial preparations, poses a significant challenge in precisely defining the pharmacological properties of this product.[6]

## Pharmacokinetics

Studies on the absorption, distribution, and metabolism of ginkgo constituents have employed the standardized extract and some individual components thought to be responsible for biological activity.[7] In general, the bioavailability of orally administered flavonoids is limited because of their low lipid solubility. Flavonoid metabolites have been identified in rats after oral administration of leaf extract. These include 4-hydroxybenzoic acid conjugate, 3-methoxy-4-hydroxybenzoic acid, hippuric acid, and 4-hydroxyhippuric acid.[8] While some investigators report that no intact flavonoids appear in rat or human blood after oral administration of the standardized ginkgo extract, others have detected quercetin, kaempferol, and isorhamnetin/tamarixetin in rat blood and brain following ingestion.[9] These flavonoids have also been reported to be present in the hippocampus following administration of the extract. Continued oral administration to rats is reported to increase the brain accumulation of these substances.[9]

The different findings with regard to flavonoid bioavailability in rats could be attributable to differences in the administered doses or to the sensitivity of the analytical procedures employed to identify these substances in blood. In general, however, it appears that the ginkgo flavonoids are extensively metabolized in the gastrointestinal tract following ingestion, primarily to phenolic acids. As the extent of this metabolism appears to be greater in humans than rats, the types of flavonoid metabolites detected in the blood and urine following oral administration differ between the two species.[7] Intravenous administration to rats of the standardized extract reveals that when the ginkgo flavonoids are placed directly into the bloodstream the half-lives of kaempferol and isorahamnetin are less than two hours, and for quercetin, slightly less than four hours.[10]

Taken together, these findings suggest that very little, if any, of the ginkgo extract flavonoids reach the bloodstream unchanged following oral administration in humans and that, even if they did, their biological half-lives are relatively short. These pharmacokinetic data indicate it may be inappropriate to extrapolate ginkgo extract results from in vivo rodent studies to the clinical situation. Moreover, these findings suggest that any therapeutic response to the flavonoid components of ginkgo extract is due to actions of their metabolites rather than to the parent compound in the leaf. This is important when interpreting the results of experiments aimed at defining the pharmacodynamics of ginkgo constituents, as it would appear to be more appropriate to study responses to the relevant flavonoid metabolites rather than to extract constituents themselves.

The absorption characteristics of ginkgo terpene trilactones are quite different from the flavonoids.[7] Rat and human studies indicate that the vast majority of the ingested ginkgolides A, B, and bilobalide are readily absorbed from the gastrointestinal system following oral administration of the standardized ginkgo extract. In a study with human volunteers, the time to reach the maximum plasma concentration of ginkgolides following administration of dried ginkgo leaf extract was two hours for all of these ingredients, with the elimination half-lives being approximately 2.5 hours for each.[11] Very little ginkgolide C appears in blood following its oral administration. In addition, measurable quantities of ginkgolides A, B, and bilobalide are detectable in a rat brain after a single oral administration of the standardized extract.[12] Thus, unlike the flavonoids, it appears that the ginkgo terpene trilactones may penetrate into the human brain following oral administration.

A number of studies were undertaken to determine the effect of ginkgo on the metabolism of other drugs. Because ginkgo may be taken for extended periods by individuals also consuming prescription or over-the-counter medications, it is important to know whether the extract might modify the breakdown of these other agents and thereby increase or decrease their blood levels and therefore their effectiveness. The most common studies of such interactions involve in vitro examinations of the ginkgo extract, or some of its known chemical constituents, on enzyme activity in human or laboratory animal tissue, or in cell systems containing a particular drug metabolizing enzyme. In general, depending on its concentration, ginkgo extract may increase or decrease the activities of various drug metabolizing enzymes.[13,14] It is reported that a particular human liver enzyme is inhibited by the ginkgo extract and that this effect is probably not mediated by either the terpene trilactones or flavone glycosides. While the flavone aglycones resulting from metabolism of the parent compound in the extract inhibit this enzyme activity, it was suggested the concentrations needed for this may be higher than those achieved following oral administration of the extract to humans. The metabolizing enzyme inhibition might therefore be due to the presence of some other, perhaps as yet unidentified, substance in the ginkgo powder.

It has also been reported that the activity of other drug metabolizing enzymes are enhanced or inhibited by certain concentrations of ginkgo extract.[13] The possible clinical significance of such findings is illustrated by the report that ginkgo administration significantly enhances the metabolism of omeprazole, an ulcer medication, in humans.[15] This suggests that ginkgo ingestion could reduce or enhance the clinical potency of some prescription medications.

While it appears ginkgo extract has the potential to influence the metabolism, and therefore the response, to a number of drugs, the degree of risk for such interactions is unknown as most of the studies addressing this question were conducted in vitro. Because such experiments seldom reveal the extract constituent responsible for any observed effect, it is impossible to know whether a sufficient amount of the relevant component is absorbed from the orally ingested product to have such an effect on drug metabolism in vivo. Nonetheless, caution should be exercised when consuming ginkgo extract with conventional medications, as the former might affect the response to the latter.

## Pharmacodynamics

Over the past three decades scores of in vitro and in vivo animal studies have been performed to determine the mechanism of any therapeutic action attributed to ginkgo extract.[7] The results suggest that the extract itself, or selected individual chemical constituents, can influence virtually all brain neurotransmitter systems, depending on the concentration or dose examined. Given the variety of tests employed, possible differences in the chemical composition of the extracts, and variations in experimental conditions, it is not surprising that these results are often conflicting. For example, while some show that the standardized ginkgo extract inhibits norepinephrine, dopamine, and serotonin uptake into rat brain neurons, others report the extract enhances the accumulation of these neurotransmitters. Work has also suggested the extract, or certain constituents, are capable of modifying neurotransmitter metabolism, the accumulation of neurotransmitter precursors, and the number of neurotransmitter receptors in the brain. Again, it has not been possible to determine whether such effects occur in humans following consumption of the standard quantity of ginkgo extract.

In vitro studies also suggest that ginkgo extract can slow apoptosis, or cell death, presumably because flavonoids are thought to scavenge free radicals and other reactive oxygen species. However, studies indicate that flavonoid antioxidant activity is unlikely to occur in vivo given the concentrations of these substances absorbed after oral administration, and their rapid and extensive metabolism in the body. Although some have suggested that the terpenoid constituents of the extract also inhibit cell death, others report no effect in this regard.[7]

Because the in vitro laboratory studies suggest that the flavonoid components of ginkgo may reduce the levels of reactive oxygen species, it was reasoned these compounds might be of benefit in treating neurodegenerative and cardiovascular disorders, as well as age-associated neuronal cell loss.[16,17] It is unknown, however, whether in vitro rodent brain studies supporting this notion are of relevance with regard to any clinical response given the aforementioned differences in the extent of metabolism of ginko extract between humans and rats, and the fact that it is unlikely these flavonoids are absorbed intact. Accordingly, for a pharmacologist it is difficult to know which, if any, of the effects found with these in vitro studies could be anticipated in humans. There is also uncertainty as to whether the amount of extract used for the in vitro work, or the doses employed for in vivo laboratory animal studies, approximate the concentrations of these extract constituents or metabolites present in the human brain following oral administration of the commercial product.

It has been known for some time that the terpene trilactones are platelet-activating factor (PAF) receptor antagonists.[18,19] As PAF is a potent, endogenous mediator of bronchoconstriction and platelet aggregation, blockade of its action could explain the purported beneficial effects of ginkgo extract in the treatment of asthma, other inflammatory conditions, and clotting disorders. However, the ginkgolides appear to have only modest affinity for the PAF receptor, suggesting they are weak in this regard. Nonetheless, it has been shown that administration of ginkgo extract to humans decreases erythrocyte aggregation and increases blood flow in the nail fold capillaries.[20] It has also been reported that the duration of the PAF blockade of an inflammatory response in human skin is brief following oral administration of a ginkgolide preparation.[18] This is consistent with the pharmacokinetic studies indicating the elimination half-lives for the ginkgolides in humans is quite short.[11] It is unknown what direct effect PAF receptor antagonism has on central nervous system activity.

The effects of ginkgo extract, and in particular the terpene trilactones, on animal behavior have been examined. These substances are reported to both enhance and reduce anxiety, to display an antidepressant effect, and to show promise as a possible treatment for drug abuse. The ginkgolides are also reported to enhance vigilance and alertness, suggesting an activating effect on the central nervous system.[7] Again, determining the possible clinical relevance of these data is problematic because of the differences in doses and in the metabolism of extract constituents between humans and laboratory animals. Nonetheless, such laboratory findings stimulated a host of studies aimed at defining the effect of ginkgo extract on a number of clinical parameters, including cerebral blood flow and alertness, and on conditions such as age-related cognitive decline and Alzheimer's disease.[7] Included among these are several double-blind, placebo-controlled studies aimed at determining whether ginkgo extract is of benefit in enhancing memory and cognition, especially in those with neurodegenerative disorders. Such studies are particularly challenging because of the difficulty in objectively measuring changes in these mental attributes over time. To this end, rating scales are used to quantify alterations in various parameters as described by the subjects, their caretakers, or clinical personnel.

A meta-analysis is a retrospective study combining and analyzing the results of a large number of previously published experiments. By combining such data from many clinical studies, the number of observations is increased, as is the likelihood of detecting a small, but clinically significant, effect. In conducting a meta-analysis it is important that all studies included be well-controlled and involved a similar patient population. The results of a meta-analysis covering 36 clinical trials of ginkgo extract have been reported. All the studies chosen for this analysis were randomized, double-blind, placebo-controlled trials. Trials with Alzheimer's patients were included as well. The results indicate no consistent clinical benefit associated with the consumption of ginkgo leaf extract.

Recently completed prospective clinical studies include one involving 118 cognitively intact subjects, 85 years of age or older, who took ginkgo leaf extract for 42 months. The aim of this trial was to determine whether ginkgo improves cognition in the elderly. No significant difference in the rate of cognitive decline was noted between those taking ginkgo and the control subjects.[21]

Another clinical trial was the Ginkgo Evaluation of Memory (GEM) study. This undertaking, funded by the United States National Institutes of Health, involved five academic medical centers and 3,069 individuals aged 72 to 96 years. Over a six year period these individuals participated in a double-blind, randomized study of ginkgo to assess its effects on memory and on the incidence of Alzheimer's disease. The subjects received either placebo or 120 mg of leaf extract twice daily. Several different rating scales were used for evaluation of mental state, memory function, and development of symptoms associated with Alzheimer's disease. No significant differences were noted for any of these measures between the treated and control groups, suggesting that ginkgo leaf extract is ineffective for delaying memory decline in the elderly and for slowing or preventing the onset of Alzheimer's disease.[22,23]

Attempts have been made to assess the effect of ginkgo leaf extract in combination with standard medications used to treat psychiatric disorders. In one study, the extract was taken by treatment-resistant schizophrenic patients along with clozapine, an antipsychotic agent. For this trial, 42 patients received, in combination with clozapine, either ginkgo extract or placebo for 17 weeks. While the addition of the ginkgo extract may have enhanced the effectiveness of clozapine in reducing the negative symptoms of this disorder, there was no significant improvement in overall symptomatology.[24]

While there have been dozens of published reports suggesting ginkgo extract has positive effects in enhancing cognition and memory in humans, and there have been laboratory animal studies suggesting it may delay the progression of neurodegenerative disorders, well-controlled, large-scale clinical trials have failed to confirm these findings. Accordingly, from a pharmacological perspective, the utility of ginkgo extract in treating central nervous system disorders remains unproven. This conclusion supports the hypothesis that any effects of ginkgo extract on the central nervous system are minor, subtle, or evident only in certain types of individuals.

## Adverse Effects

Side effects to the recommended doses of ginkgo extract appear minimal and are quantitatively and qualitatively similar to complaints reported after consumption of a placebo. This safety profile explains, in part, the popularity of this product. Side effects include gastrointestinal discomfort, nausea, diarrhea, and headaches. Some may experience allergic reactions as ginkgo leaves contain allergens related to those found in poison ivy. Because constituents of the extract may inhibit clotting, it is possible that consumption of the powder could lead to bleeding, especially in those taking anticoagulant drugs. While clinical studies on this question are inconclusive, caution should be exercised by those taking drugs known to slow blood coagulation, such as aspirin and warfarin.[25] Because there are reports suggesting the constituents of ginkgo extract might inhibit or activate drug metabolizing enzymes, consumers should be alert to possible changes in the responsiveness to prescription or over-the-counter medications when taking ginkgo products.

As the precise mechanism of action of any central nervous system effects of ginkgo is unknown, it is difficult to speculate about the possibility of adverse consequences due to pharmacodynamic interactions with known psychotherapeutics. Nonetheless, it is recommended that ginkgo extract should not routinely be taken by those on antidepressant medications unless under the close supervision of a physician.

## Pharmacological Perspective

Given the popularity of ginkgo extract, many obviously feel they benefit from the consumption of this herbal product. While its lack of side effects is a positive characteristic, it might also be interpreted as indicating the product has little, if any, biological activity. Indeed, from a pharmacodynamic standpoint, it is difficult to take seriously the number of neurochemical responses that have been attributed to ginkgo extract and extract constituents from laboratory animal studies as such generalized actions would be expected to result in a number of serious side effects, none of which are observed in clinical trials.

The pharmacological data do, however, suggest the possibility that ginkgo leaf extract might diminish the inflammatory and clotting responses to PAF, and therefore symptoms of some clinical conditions. Its value as an agent for treating central nervous system conditions remains unproven, at least with respect to enhancing cognition, slowing cognitive decline, or delaying the onset of Alzheimer's disease. In the absence of large scale, placebo-controlled clinical trials for the treatment of other central nervous system disorders, it is impossible to draw any conclusions about the overall effectiveness of this product. Inasmuch as ginkgo leaf extract has been widely available for years, and taken by millions of people for a variety of reasons, it would be anticipated that any obvious and consistent effect on central nervous system activity would be known by now. The possibility remains that ginkgo constituents may influence brain function in a select group of individuals experiencing certain types of central nervous system dysfunction. However, until it is possible to identify which, if any, components of the ginkgo extract, or their metabolites, penetrate into the human brain at concentrations sufficient to influence neuronal activity, the pharmacological properties, and therefore the therapeutic potential, of this product will remain undefined.

# 6

## St. John's Wort (Hypericum perforatum)

Classified as a mood disorder, depression was historically included under the general heading of melancholia. Feelings of depression, which are experienced by everyone, most often occur in response to some external event, such as the death of a loved one, as a result of a rejection, or because of a professional setback, such as a job loss. This condition, situational depression, is characterized by feelings of sadness, gloom, and frustration. Over time these symptoms abate without the need for medication or the assistance of a mental health professional. The support and understanding of family and friends, and the passage of time, are usually sufficient for overcoming situational depression.

Clinical depression, sometimes referred to as major or endogenous depression, is a severe, life-threatening disorder for which there is no apparent precipitating cause. The symptoms may include lethargy, sleep disturbance, changes in appetite, despondency, morbid thoughts, feelings of worthlessness, anhedonia, which is an inability to experience pleasure from activities that were once considered enjoyable, and thoughts of suicide. The degree and duration of these symptoms far exceed those normally associated with any external event and cannot be traced to any underlying somatic condition. Depending on the symptomatology, clinical depression is characterized as mild, moderate, or severe. It is often associated with other mental disorders such as generalized anxiety and psychosis.

While major depression has been considered a distinct clinical entity only since the mid-nineteenth century, a treatise published 200 years earlier lists 14 different herbs that could be used to "drive away sorrow and increase joy in the mind."[1] Before the seventeenth century, melancholy was thought to be caused by an excess of black bile, one of the four humors of Hippocrates. As black bile was associated with cold and winter, herbs considered to be warming were recommended as treatments for conditions believed to result from an imbalance of this humor. As a warming herb, St. John's wort was a candidate therapy. Today St. John's wort is one of the more popular herbal supplements, especially in Europe where consumers spend billions of dollars each year on this plant extract as a treatment for feelings of depression.

In spite of its popularity and centuries of use, the antidepressant value of St. John's wort remains controversial. The Roman scholar Pliny wrote 2,000 years ago that St. John's wort is of benefit in the treatment of hysteria, a form of melancholia.[2] Such reports cannot be taken as proof of efficacy for treating clinical depression as it is defined today because, in part, of the differences in the reported length of treatment needed for a response. Thus, the ancients believed that herbal remedies acted quickly in treating mood disorders. However, prescription antidepressants typically require days or weeks of continuous administration before improvements are noted in the subjective symptoms of this disorder. This delayed response is thought to be due to the need for drug-induced adaptations in brain structure and neurochemistry. It is possible that the difference in the time to response to modern medications and the ancient Roman remedy is due to the fact that no distinction was made in the first century between those suffering from situational and major depression. Also, because mood and affect are dramatically influenced by expectations, the placebo response rate is high in depressed patients, making it difficult to discern the extent to which a drug or herbal remedy speeds recovery as compared to the influence of psychological factors and the normal healing process. Even today this concern has some questioning the clinical value of prescription antidepressants, let alone herbal products. Accordingly, it is difficult to prove scientifically that a plant extract is of any benefit as a treatment for depression, whether it be situational or clinical. As the published pharmacological data are conflicting, the value of St. John's wort as an antidepressant remains uncertain.

## Botany

St. John's wort (Hypericum perforatum) is a perennial herb that first appeared in Western Europe but is now widely distributed throughout temperate regions of the globe, including North America. The name *Hypericum* is a combination of the Greek *hyper*, meaning above, and *eikon*, an image or icon. This refers to the Greco-Roman custom of placing Hypericum branches above statues to ward off evil spirits. The term *perforatum* relates to the perforated appearance of the leaf that is due to the presence of glands along its surface. The common English name refers to St. John the Baptist, whose birthday coincides with the time of year, early summer, when this plant begins to flower. In the ancient world festivals were commonly held during this period to celebrate the beginning of summer and, thanks in part to the flowering of St. John's wort, the flight of evil spirits.

Hypericum perforatum is a hybrid of two wild plants, Hypericum maculatum and Hypericum attenuatum. As a result of this cross, the sterile hybrid Hypericum perforatum has 32 chromosomes, double the number found in the two parent plants. Siberia is believed to be the location of the cross as it is here that the distribution of the more Western maculatum overlaps the more Eastern attenuatum. A perennial herb, the aerial portion of the plant dies each year, emerging from the roots again each spring.

St. John's wort is a member of the family Clusiaceae (formerly Guttiferae). Gray's Botany lists 25 different species of Hypericum that are found in the eastern United States. In North America St. John's wort grows wild from Canada to the southern United States. The five-petalled, yellow flowers blossom from June throughout the summer. The plant is commonly seen in vacant lots and along roadsides. It is considered a weed in cultivated fields.

Ascyrum, or St. Peter's wort, is the only other genus of the family Clusiaceae that grows wild in North America. The English botanist John Gerard wrote that St. Peter's wort has therapeutic effects similar to St. John's wort.

Because of the growing interest in St. John's wort, the chemical ingredients of several species of Hypericum are being studied as herbal remedies. As with all plants, there are variations in the constituents, and their relative quantities, among different species and subspecies of Hypericum. Moreover, during the spring growing period the concentrations of chemical constituents generally increase, reaching their peak at flowering. While the commercial St. John's wort preparation, the dried residue of an alcohol extract, is standardized to 0.3% hypericin, the relative quantities of other chemical constituents can vary widely depending on the species and subspecies, its geographical location, the local weather conditions for that year, and the seasonal period of collection. This makes it imperative to identify the plant constituent, or constituents, most responsible for influencing central nervous system function to define the pharmacological characteristics, and therefore the therapeutic potential, of this extract.

## Therapeutic Uses

St. John's wort has long been touted as a remedy for a variety of conditions. A multiplicity of possible uses is not unique to it, or to herbal supplements, as a group. Indeed, many contemporary prescription drugs are employed for multiple purposes. For example, morphine, a component of the opium poppy, Papaver somniferum, has for centuries been employed as an analgesic and as a treatment for diarrhea. Hypericum, which is used most commonly today as a treatment for depression, is listed in a sixteenth century Chinese text on medical herbs as a remedy for "miasmatic disease," or malaria.[3] Likewise, early Europeans employed Hypericum mainly as a treatment for malaria and to relieve sciatic pain, common conditions of the time.[4] This is known from the *Medicina Antiqua* (Codex Vindobonensis 93), an extant ninth century copy of the fourth century Pseudo-Apuleius manuscript that describes a drink made from Hypericum leaves and seeds. The same recommendations are found in seventeenth century publications. Gerard stated that if consumed for 40 days the Hypericum seed cures sciatica, as well as tertian and quartan fevers (malaria).

St. John's wort has also in the past been considered a treatment of choice for burns, wounds, and skin ulcerations. The preparation used for these purposes was an olive oil extract of the pressed leaves and flowers. A sixteenth century medical herb volume authored by the Jesuati Friars of Lucca, Italy, included a balsam that was said to heal any kind of wound within 24 hours. A balsam is a scented plant product. Referred to as precious oil, this balsam was composed primarily of the leaves, flowers, and seeds of St. John's wort, pounded and mixed with olive oil. The mixture was recommended for its value in healing bruised or damaged skin.[5]

Today, the most common use for St. John's wort is as an antidepressant. This is a relatively recent development, as illustrated by the fact that the classic 1931 M. Grieve text on herbal remedies lists 15 uses for St. John's wort, only two of which were for the treatment of central nervous system disorders: hysteria and nervous depression.[6] By 1995 the interest in St. John's wort as an antidepressant had grown sufficiently to warrant a review of this topic.[7] Since then there have been many reports on its possible antidepressant properties, especially for the treatment of mild to moderate forms of major depression. There is as well a growing interest in isolating and identifying the chemical constituents in St. John's wort extracts that could be responsible for any perceived therapeutic benefit, including possible anti-inflammatory, analgesic, antineoplastic, and antimicrobial effects. To obtain convincing data for any of these actions, it is critical to first identify the most pharmacologically active chemical ingredients contained in this plant extract.

## Constituents

The dried flowering tops of St. John's wort are the portion of the plant from which an extract is prepared for the treatment of depression. Among the various compounds contained in the yellow flowers are the hypericins, which are chemicals unique to the genus Hypericum. The flower is the preferred source of hypericins as the roots, stems, and leaves of St. John's wort contain only small quantities of these compounds. The hypericins are produced in the oil glands located on the tips of the flower shoots and in the petals. Because this molecule is a chromophore, the hypericins are responsible for the red coloration of the dried flowers and for the photosensitivity that develops in livestock that consume this plant.

Chemically, hypericin and related compounds are condensed anthrones. Hypericin itself is a napthodianthrone or bianthraquinone. That is, it is composed of two anthraquinones joined by three phenyl-phenyl linkages. This structure is sometimes humorously referred to as "hydroxy-chickenwire" because of its complexity (see Figure 6.1).

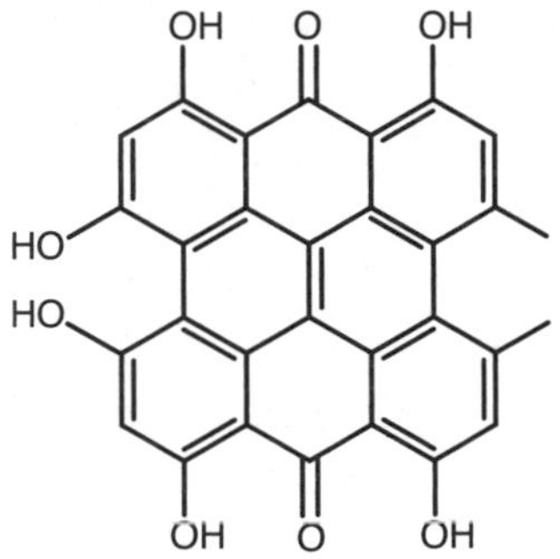

Figure 6.1 Chemical structure of hypericin (Wikipedia)

Hyperforin, a phloroglucinol, is another pharmacologically interesting chemical found in the flowering tops of St. John's wort (see Figure 6.2). The highest concentrations of hyperforin are in the maturing seed capsules, with the flower content increasing during maturation. Of all the chemicals isolated from St. John's wort, the evidence is greatest for hyperforin as the most important for the purported antidepressant and mood-elevating activities. The concentration of hypericin in preparations of the flowering tops of St. John's wort is standardized at 0.3%. While the hyperforin content is more variable, it averages about 4% in most commercial extracts.

Besides its potential antidepressant activity, at the proper concentrations hyperforin is reported to inhibit the growth of some bacteria and viruses.[8] Such antibiotic activity could explain why St. John's wort

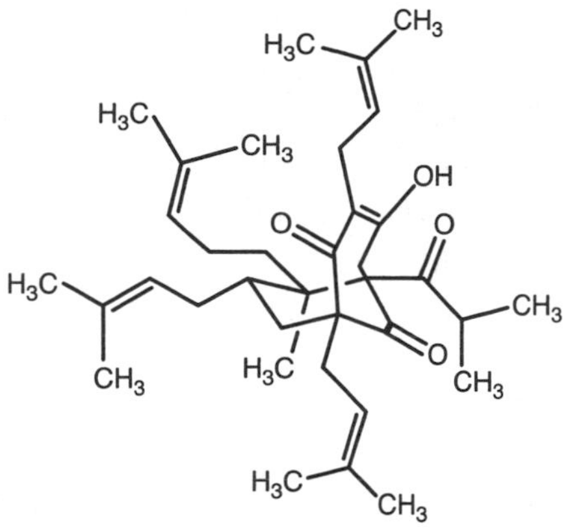

Figure 6.2 Chemical structure of hyperforin (Wikipedia)

was for centuries recommended as a wound treatment. Hyperforin also reportedly displays anti-inflammatory activity.

Other chemical constituents of St. John's wort include flavonols, tannins, and a volatile oil. These compounds are found primarily in the leaves. Flavonols, which are defensive compounds produced by many plants, protect the vegetation from ultraviolet radiation. Kaempferol and quercetin are two flavonols isolated from the leaves of St. John's wort. Tannins are complex phenols present in the stems, leaves, and flowers of St. John's wort. Like flavonols, tannins are widely distributed throughout the plant kingdom. Tannins are commonly used to treat burns and wounds because they interact with cellular proteins to form a protective layer on damaged skin. This action may underlie the beneficial effects reported by the ancients for the topical use of St. John's wort.

The translucent glands on the leaves of St. John's wort contain an essential oil that is composed of various pleasant-smelling terpenes, among which are alpha-pinene, beta-myrcene, and beta-caryophyllene. These compounds emit the odor of pine, moss, and cloves, respectively.

Given the current interest in employing St. John's wort for improving mood, significant effort has been made to identify the chemical constituents in the plant extract that may be responsible for this action. These studies include both laboratory and clinical experiments with purified substances isolated from the extract. The laboratory animal experiments focus particularly on determining whether extracts of St. John's wort, or individual agents derived from them, influence brain neurotransmitters in a manner similar to prescription antidepressants. The results of such experiments have little meaning, however, unless it can be demonstrated that pharmacologically active constituents of St. John's wort extract are absorbed in humans following oral administration and that they penetrate into the brain.

## Pharmacokinetics

Oral consumption of 900 mg of Hypericum extract by human volunteers results six hours later in a maximum plasma hypericin concentration of 7.2 ug/l, with a median half-life of distribution of six hours and a median half-life of excretion of 43.1 hours.[9] In a separate study, a 900 mg oral dose of a Hypericum extract resulted three hours later in a maximum plasma hyperforin concentration of 437.3 ug/l, a three hour median half-life of distribution and a median half-life of excretion of 8.65 hr.[10] The steady state plasma concentration of hyperforin was 180 nM.[11] These studies indicate that both hypericin and hyperforin are absorbed intact from the intestinal tract following oral administration of the extract. The differences in plasma levels for these two chemicals may explain why hyperforin is thought to be most responsible for any antidepressant response to the extract. Even though the average concentration of hyperforin in the extract preparation is some 20 times higher than hypericin, these data indicate the peak plasma level of hyperforin is nearly 70 times greater than for hypericin. This, along with the shorter time to attain peak plasma levels, suggests that hyperforin is more readily and completely absorbed into the bloodstream than hypericin.

These data notwithstanding, it remains unclear whether these substances are transported into the brain in quantities sufficient to affect behavior.[12-14] Studies with human liver fractions indicate that hyperforin is converted to 57 different metabolites.[15] While metabolites are generally less likely to accumulate in the brain than parent compounds, the possibility remains that responses to St. John's wort are mediated by some, as yet unidentified, chemical constituent or its metabolite. In either case, the lack of evidence for brain accumulation of hyperforin or hypericin, let alone for the flavonols, tannins, or any other chemical known to be present in St. John's wort extract, leaves unanswered the question about which of these, or which combination of them, is responsible for its effects following oral administration. The absence of this information is a significant hindrance to characterizing a possible mechanism for the purported antidepressant effect of this extract.

## Pharmacodynamics

Clinically effective antidepressants were first identified in the 1950s. Since then a host of prescription agents have been developed for this purpose. A driving force in these drug discovery efforts are the data indicating that a common characteristic of antidepressants is that they enhance the synaptic content, and therefore activity, of certain neurotransmitters in the brain, in particular serotonin and norepinephrine. This is generally accomplished in one of two ways. Most antidepressants inhibit the reuptake of one or both of these neurotransmitters after their release from presynaptic neurons, thereby prolonging the presence of the transmitter in the synaptic cleft. Fluoxetine, or Prozac®, is an example of such an agent, with its clinical efficacy being due to selective inhibition of neuronal serotonin uptake. Another mechanism for enhancing the activity of these neuronal systems is inhibition of neurotransmitter metabolism. As serotonin and norepinephrine, as well as other neurotransmitter substances, are metabolized by monoamine oxidases, drugs that block one or more of these enzymes display antidepressant activity. Moclobemide, or Manerix®, is an example of such an agent. While older antidepressants interact with numerous neurotransmitter systems, enhancement of central norepinephrine and serotonin transmission is the most common feature of this drug class. Effects on other transmitter systems are often responsible for the side effects encountered with these agents. These findings led to the monoamine hypothesis of depression, which posits that the condition is the result of underactive norepinephrine and serotonin pathways in select brain regions. Accordingly, agents capable of enhancing the activity of these systems are antidepressant candidates.

Given this information, those interested in determining the antidepressant potential of St. John's wort extracts examine their effects on brain neurotransmitters, with particular emphasis on serotonin and norepinephrine. Administration of Hypericum extracts reportedly induces antidepressant-like alterations in neurotransmitter dynamics in rats, causing an increase in the levels of dopamine and serotonin in the brain.[16] Moreover, it was reported that hypericin inhibits monoamine oxidase, an action that would increase brain levels of norepinephrine, serotonin, and other substances.[17] However, subsequent work indicates that Hypericum extracts, and hyperforin itself, are only weak inhibitors of this enzyme. Given the concentration in the brain necessary for this effect, it seems unlikely that inhibition of monoamine oxidase is the mechanism of any antidepressant action of St. John's wort.[11]

There are also reports that hyperforin, like many clinically useful antidepressants, inhibits the neuronal uptake of norepinephrine and serotonin.[11] However, unlike established antidepressants, hyperforin inhibits the neuronal uptake of many other neurotransmitters as well, including dopamine, γ-aminobutyric acid (GABA), and glutamic acid. In rat brain tissue the hyperforin concentration necessary to inhibit significantly the neuronal accumulation of these transmitters is approximately 1 μM. However, following the 300 mg/kg dose administered to rats, the plasma steady-state concentration of hyperforin is nearly 2 μM, a value that is ten times higher than the steady-state concentration in humans. Accordingly, unless hyperforin accumulates in the human brain to concentrations that greatly exceed its plasma levels, and there is no evidence indicating this is the case, it is unlikely there are sufficient quantities in the human brain to inhibit uptake of these transmitters.

It is suggested that this nonselective effect on transmitter uptake is due to a hyperforin-induced activation of transient receptor potential channels (TRPC), in particular TRPC6, which, in turn, provokes an intracellular accumulation of sodium and calcium ions.[18,19] It is believed that the increase in intracellular sodium is responsible for the hyperforin-induced decrease in neurotransmitter uptake. There is also evidence that activation of TRP channels promotes dendritic growth.[20] As dendritic sprouting is believed to be an important component of antidepressant drug action, this finding supports the notion that hyperforin could be a critical component in St. John's wort for any antidepressant effect.

St. John's wort extracts have also been shown to activate brain sigma receptors, a site thought to be a target for antidepressants.[21,22] As stimulation of sigma-1 receptors causes neuronal firing, this action would lead to activation of brain neurotransmission. It is unclear whether a chemical in the St. John's wort extract, or a metabolite of one of its chemical constituents, is responsible for stimulating the sigma-1 site in the brain.[22]

Another possible explanation for the antidepressant activity of St. John's wort relates to a proposed mechanism for its anti-inflammatory effect. Thus, studies suggest that the anti-inflammatory response to hyperforin is due to an interaction with nitric oxide synthase. This enzyme is responsible for the formation of nitric oxide, an endogenous gas known to contribute to the inflammatory process and to serve as a neurotransmitter in the brain. As there are data suggesting that nitric oxide may be involved in the pathogensis of depression, hyperforin-induced blockade of its production could contribute to an antidepressant response.[23]

Kaempferol and quercetin, two flavonols found in the leaves of St. John's wort, are thought to inhibit 5-lipoxygenase, an enzyme that produces endogenous inflammatory substances. As these flavonols display anti-inflammatory activity in humans, the effect on 5-lipoxygenase could provide a plausible explanation for their action in this regard. As it appears that neither of these compounds enters the brain in sufficient quantities to influence neuronal activity, it is unlikely this action, or their reported ability to reduce oxygen radicals or to inhibit protein kinase C, a mediator of intracellular signaling, contributes to any central nervous system effects of St. John's wort.

Taken together, these in vitro data suggest that St. John's wort extracts in general, and hyperforin in particular, have effects on brain neurochemistry similar to those reported for antidepressant drugs. This conclusion is reinforced by in vivo animal experiments. Thus, the behavioral endpoints for conventional animal models used for screening antidepressant drug candidates generally reflect an activation of norepinephrine or serotonin pathways in the brain. Some entail the prevention or reversal of stress-induced behaviors, as depression is often associated with prolonged periods of anxiety. In one study of St. John's wort, stresslike behavioral and neurochemical changes were induced in mice by the continuous administration of corticosterone, a stress hormone, for seven weeks. Administration of Hypericum perforatum extract for three weeks reversed the steroid-induced behaviors and caused antidepressant-like changes in the dendritic spines of the dentate gyrus, an area of the brain thought to be involved in depression.[24] These findings are consistent with the in vitro neurochemical effects of St. John's wort suggesting that components of this plant extract, in particular hyperforin, may inhibit neurotransmitter accumulation and promote dendritic growth, perhaps through activation of the TRPC6 receptor.

While a great deal of emphasis has been placed on the assumption that hyperforin is the extract constituent most responsible for the antidepressant action of St. John's wort, significant questions remain about the validity of this hypothesis. As noted previously, there are no data demonstrating conclusively that sufficient quantities of hyperforin accumulate in the brain following oral administration to laboratory animals or humans, and the human plasma levels of this substance are well below what would likely be needed to achieve the brain concentrations necessary to inhibit neurotransmitter uptake or metabolism. It is also notable that Hypericum extracts lacking hyperforin display antidepressant activity in animal tests, suggesting it may not be the only active constituent in this regard.[25] To the extent the results of animal studies are taken as proof for an antidepressant effect, these findings add further to the uncertainty about which of the St. John's wort extract constituents possess antidepressant properties and, if such activity is present, the mechanism of action of the critical component.

Prospective clinical studies, as well as meta-analyses, have been conducted to determine whether St. John's wort displays antidepressant activity in humans. One meta-analysis included 15 placebo-controlled, double-blind studies involving 1,000 depressed patients.[26] The dose of the Hypericum extract varied between 300 to 1000 mg. The results indicated that patients taking Hypericum improved more quickly than those on placebo, with no greater incidence of side effects. The findings of a later meta-analysis also indicate that St. John's wort is more effective than placebo as a treatment for mild to moderate depression.[27] The daily dose of Hypericum extract in this study was 600 to 900 mg.[27] In a 2010 review of Hypericum extract clinical studies, the authors concluded that the weight of evidence supports its efficacy and tolerability as a treatment for mild to moderate depression.[28] Described in this review is the individual testing of four different Hypericum extracts, with the results indicating that all are more effective than placebo in treating some forms of depression. They also report that these extracts are better tolerated than prescription antidepressants. While most clinical trials have limited the testing of St. John's wort to the management of mild to moderate depression, the results of one meta-analysis suggested that Hypericum extracts were superior to placebo for the treatment of more severe depression as well.[29]

There have also been prospective studies suggesting that St. John's wort extract is superior to placebo and is as effective as prescription drugs for the treatment of mild to moderate depression.[30,31] However, other prospective clinical trials have failed to demonstrate any efficacy for Hypericum extract in this regard.[32,33] With regard to the latter, in the late 1990s the United States National Institutes of Health sponsored a large prospective study of Hypericum extract involving 340 patients with moderately severe major depression.[34] The response to the extract was found to be no better than placebo, suggesting no benefit for St. John's wort in the treatment of this condition. Confounding this interpretation, however, is the fact that sertraline, a prescription antidepressant included in the study as a positive control, also failed to demonstrate a significant benefit in this patient cohort. Sertraline, but not St. John's wort was, however, superior to placebo on secondary tests of efficacy. This study leaves unanswered the question as to whether St. John's wort is of benefit for those with only mild to moderate forms of depression.

The results of these clinical trials fail to demonstrate conclusively that St. John's wort is a consistently effective antidepressant. The positive results suggest it may display greatest benefit as a treatment for mild to moderate depression, while the negative trials indicate it is no better than placebo in this regard. These disparate findings are likely due to the inherent difficulties associated with clinical trials for demonstrating antidepressant efficacy because, in part, of high placebo response rates, and, in this case, the variations in the chemical compositions of the extracts employed. The latter problem, which is a feature of studies with herbal supplements, makes it difficult to compare directly the results of these clinical trials as the product being tested is not identical among them.

## Adverse Effects

While St. John's wort extracts, at the doses normally consumed, have fewer side effects than prescription antidepressants, administration can be associated with gastrointestinal problems, such as nausea and diarrhea. Overall, however, the reported incidence of side effects with St. John's wort extract appears to be no greater than that recorded for depressed patients taking a placebo.[35]

The two greatest concerns regarding possible adverse effects of St. John's wort are photosensitivity and drug interactions. As sensitivity to sunlight is known to develop in grazing cattle that consume large quantities of St. John's wort, it is possible that patients taking this extract for prolonged periods might develop a dermatitis when exposed to ultraviolet light. In general, however, the doses of St. John's wort extract typically administered to humans are well below the amount generally necessary to cause photosensitivity.

Human consumption of St. John's wort extract increases the production of certain drug metabolizing enzymes and other proteins responsible for transporting drugs throughout the body.[36] These effects are documented in a study involving healthy male volunteers who were administered Hypericum extract for 14 days. Following this treatment there was a 1.4-fold increase in the levels of some gastrointestinal drug transporters and in certain drug metabolizing enzymes. This would explain the reported decrease in the plasma concentrations of several commonly used drugs when they are taken in combination with St. John's wort. Among these are digoxin, a treatment for heart failure, amitriptyline, an antidepressant, alprazolam, an anxiolytic, dextromethorphan, a cough suppressant, and cyclosporine, an immune system modulator.[37] It is believed that hyperforin is primarily responsible for altering the levels of the drug metabolizing enzymes.[38] These findings underscore the importance of being cautious about taking St. John's wort while on other medications, as the extract could decrease the effectiveness of the prescription drugs.

While it has been suggested that St. John's wort extract poses little risk to the fetus, it may not be wise to use this product during pregnancy.[39] In this regard, it has been reported that, in vitro at least, both St. John's wort and hypericin increase placental transport of calcium. Such an effect could have adverse consequences for the fetus if it occurs in vivo.[40]

## Pharmacological Perspective

St. John's wort is a plant of considerable interest because of its unique chemical constituents and the biological activity associated with some of these compounds. The fact that St. John's wort has been used since antiquity for medicinal purposes suggests that at least some of its beneficial effects are empirically obvious. However, just because an action was noted and recorded by our ancestors does not guarantee it will be proven true when tested using rigorous, modern scientific methods. Take for example the ancient belief that St. John's wort chases away evil spirits. Even though the scientific evidence supporting its efficacy as an antidepressant is greater than for its value as a talisman, its effectiveness as a treatment for depression remains uncertain. This is due in large measure to the mixed results from clinical trials, especially the prospective studies. There is also uncertainty regarding the extent to which any of the constituents of St. John's wort extract, in particular hyperforin, penetrate into the brain at concentrations sufficient to influence chemical neurotransmission. Moreover, many find it difficult to accept the proposal that the antidepressant action of hyperforin is related to its nonselective inhibition of neurotransmitter uptake because such an action should be associated with a host of central nervous system side effects. As no such side effects are observed at the doses of St. John's wort normally administered, either this effect on neurotransmitter uptake does not occur in humans, or the quantities of hyperforin that accumulate in the brain are insufficient for influencing these transmitter systems. Thus, the safety of St. John's wort is an argument against any significant central nervous system effect, including an antidepressant action. What does seem certain, however, is the potential for St. John's wort to influence the response to prescription drugs by enhancing their metabolism and elimination from the body. This makes it imperative that the risk/benefit ratio of consuming this product be taken into account by those receiving medications for other conditions. Given the interest in this plant extract, work will undoubtedly continue to more clearly define its clinical utility. The weight of evidence to date suggests this may include anti-inflammatory and antimicrobial, if not antidepressant, activities.

# 7

# Valerian (Valeriana officinalis)

Various rituals and remedies are employed for treating anxiety and insomnia. Among these are warm milk, soft music, soothing background noise, yoga, and aroma therapy. However, in centuries past the most popular approach was the consumption of agents believed to relieve tension, including valerian. The historical interest in calming agents is evidenced by the immediate, and ongoing, popularity of fermented alcoholic beverages, which were first produced in China and the Mideast 9,000 years ago. The more potent distilled spirits came later, with the arrival of alchemy in southern Italy during the twelfth century AD. Ethyl alcohol, or ethanol, the pharmacologically active constituent of these beverages, belongs to a drug class known as central nervous system depressants. By reducing brain activity, such agents decrease anxiety (anxiolytic effect) and induce a feeling of drowsiness (hypnotic effect). Besides alcohol, other central nervous system depressants are the barbiturates, such as phenobarbital and thiopental, and the benzodiazepines, as exemplified by chlordiazepoxide (Librium®) and diazepam (Valium®).

The popularity of valerian root extract is linked directly to its perceived efficacy as an anxiolytic and hypnotic. Evidence of its utility in this regard can be traced to the dawn of man, with archeologists finding remnants of the valerian plant in southern European caves formerly occupied by Cro-Magnon, the earliest Homo sapiens. This indicates that ancient humans consumed this plant, no doubt for its nutritional value, but also perhaps for its calming properties.

Written records prove the medicinal properties of valerian were appreciated by ancient civilizations. Mentioned by the Greek physician Hippocrates and, later, by Galen, a Roman physician and scientist, this herb is described as a treatment for insomnia. Although valerian has for thousands of years been used to treat many conditions, its use as an anxiolytic and hypnotic has withstood the test of time. Because its calming action has been appreciated for millennia, it seems likely that constituents of valerian noticeably affect brain function when consumed in sufficient quantities. A pharmacological analysis of the published reports tends to support this conclusion.

## Botany

There are more than 200 species of valerian, many of which are not used for medicinal purposes. The most popular for herbal preparations is Valeriana officinalis. Its horizontal roots (rhizomes) and rootlets are the portions used for preparing the herbal extracts. A member of the family Valerianaceae, this species is native to Europe and western Asia. It was subsequently introduced and naturalized in temperate regions of North America.

Valerian is a perennial herb with white or pinkish flowers. The main stalk grows to about three feet and is topped with clusters of delicate, bell-shaped flowers. It has long, cylindrical roots with numerous rootlets that grow along the ground, making them readily accessible for harvesting. Valerian leaves are large and pinnate, meaning they are divided into leaflets on either side of a central axis. The term *pinnate* is derived from *pinna*, Latin for feather, which aptly describes the leaf morphology.

Two other members of the Valerianaceae used to prepare remedies are Valeriana celtica (French spikenard) and Nardostachys jatamansi (spikenard). The suffix "nard" suggests these species were, like lavender, associated with Naarda, an ancient Syrian city. Besides its use as a medicinal, extracts of spikenard rhizomes have been used in perfumes and for incense. Various species of spikenard are also native to Eurasia, India, and the Far East. Like many medicinal plants, spikenard was adapted to cultivation in the Fertile Crescent, an area that stretches from Egypt through Anatolia in Turkey to the Euphrates and Tigris river valleys in Mesopotamia, a region that today encompasses parts of Iran, Iraq, Syria, and Turkey. As the Tigris and Euphrates are two of the four rivers described in Genesis as originating in the Garden of Eden, Islamic tradition holds that spikenard may have been the forbidden fruit consumed by Adam. A seventh century BC list of Assyrian herbals includes both French spikenard and spikenard as remedies.

The various species of valerian yield somewhat different chemical constituents. None of the four Valeriana species native to North America are used for medicinal purposes. The American group includes Valeriana ciliata, a prairie species, Valeriana septentrionalis, a northern species, Valeriana uliginosa, found in northern marshes, and Valeriana pauciflora, a woodland plant. For this reason, studies aimed at defining the medicinal value of Valeriana, and the chemical constituents responsible for any pharmacological effects, generally utilize plants originally found outside of North America.

## Therapeutic Uses

A sixteenth century Chinese text on herbal medications lists spikenard as one of five odorous plants and notes that in India it was used to treat hysteria and epilepsy.[1] Over the centuries, valerian root has been employed for the treatment for many conditions. In the sixteenth century Nicholas Culpeper, an English botanist and herbalist, published a recipe for valerian root boiled with licorice, raisins, and aniseed to be used for cough and as a treatment for plague. He also recommended a tincture of valerian root for "nervous affections" including heart palpitations, vapours, and hysteria.[2] Other purported uses are for indigestion, nervous headache, muscle tension, irritable bowel syndrome, pain, and inflammation.

Valerian root is most popular as a treatment for insomnia and anxiety. Approximately one gram of the dried root, or a medicinal tea extract, is consumed for these purposes. The tea is prepared by soaking three to five grams of the dried root in hot water. An alcoholic extract (tincture) is also prepared. Given the different properties of alcohol and water, the chemical constituents extracted into the tea and tincture differ, even when they are prepared from the same root sample.

Both standardized and nonstandardized preparations of valerian, in capsules or tablets, are available for purchase, as are unprocessed samples of the dried root. The standardized preparation typically lists the calculated percentage of valerenic or valeric acids in the product. As these constituents represent only a small fraction of the chemicals in the mixture, it is difficult to know whether any pharmacological effects are due primarily to the presence of these acids, other components of the extract, or to a combination of chemicals. In some products valerian is purposely mixed with other herbals, further complicating a systematic pharmacological assessment of these products.

Some contend that to be fully effective valerian must be taken continuously for several weeks, while others report that symptom relief is almost immediate. Valerian is generally not recommended for continual use, but rather as an occasional treatment for insomnia and anxiety.

Valerian sesquiterpenes have also been reported to display anti-inflammatory activity.[3,4] This action is thought to be due to an interaction with nuclear factor kappa B, a macrophage component involved in regulation of inflammation, immunity, and cancer. Such findings have led to an interest in chemically synthesizing valerenic acid derivatives to optimize the anti-inflammatory activity of this compound. The extent to which such chemical analogs may influence brain function is unknown.

In an extensive survey of the phytotherapeutic potential of 1,000 plants, nine were considered to have the greatest therapeutic potential. Valeriana officinalis was one of these nine, suggesting it is generally believed to display noticeable pharmacologic activity.[5]

## Constituents

One of the more obvious characteristics of valerian is the odor emitted by its roots. This aroma is derived from a volatile oil produced by the plant. John Gerard, the sixteenth century English herbalist, described the scent of the fresh root as pleasant, and, 2,000 years earlier, the Greek botanist Theophrastus included valerian root as an ingredient of perfume.[6,7] The dried root, on the other hand, which is typically used for medicinal purposes, emits an odor that is unattractive to humans but apparently pleasant to rodents and cats. Indeed, legend has it that dried valerian root was used by the Pied Piper to lure the rats from Hamelin. The scent results from the chemical transformation of volatile substances during the drying process.

The fresh Valeriana root contains up to 2% volatile oil, of which bornyl acetate, a terpene, is a major constituent. The odor of the fresh root has been ascribed to this substance. The unusual odor associated with the dried root is the result of the formation of isovaleric acid by the enzymatic oxidation of bornyl acetate during drying. The aroma of isovaleric acid is variously characterized as rancid, disagreeable, or cheeselike.

As for its medicinal use, the essential oil in the root of Valeriana officinalis contains a mixture of monoterpenes, known collectively as valepotriates, and sesquiterpenes, in particular valerenic acid, which is thought to be one of the more pharmacologically active constituents of this plant. The valepotriates are so named because the monoterpene mixture contains **val**erian **epo**xy **tri**esters. The valepotriates are water-insoluble and chemically unstable. Their concentrations vary with the valerian species, the local environment, and the conditions under which the root is dried. Studies of the volatile oil taken from a single cultivar of Valeriana revealed that the concentrations of the valepotriates and the sesquiterpenes vary as much as 4-fold.[8] Others report differences of up to 600-fold in the concentrations of valerenic acid in standardized commercial preparations of valerian. These large differences in the concentrations of this substance are undoubtedly responsible, at least in part, for the variations in the biochemical and clinical responses to this herbal supplement.[9,10]

The method of extraction from valerian roots also affects the chemical composition of the final product.[11] The number and varying amounts of chemicals in the extracts make it difficult to determine all of the active principles and their possible interactions. It is impossible, therefore, for a consumer to know precisely the amount of active constituents being taken when ingesting a valerian product.

Both the valepotriates and the sesquiterpenes are thought to contribute to the sedative and anxiolytic properties of valerian root. Because of their chemical instability, the valepotriates probably do not appear in the brain following systemic administration. However, they are converted in the body to other chemicals, such as homobaldrinal, which may be biologically active in vivo.[11]

The sesquiterpene valerenic acid has received considerable attention as a pharmacologically active component of valerian. Aqueous extracts of valerian root also contain GABA, an inhibitory neurotransmitter. It has been proposed the presence of this amino acid may contribute to the hypnotic action of valerian extract.[11] This is unlikely, however, as GABA is not absorbed from the gastrointestinal tract and does not readily cross into the brain from the systemic circulation. Other chemicals identified in valerian include the lignin hydroxpinoresinol, the alkaloids actinidine, chatinine, shyanthine, valerianine, and valerine, and the flavonoids hesperidin, 6-methylapigenin, and linarin. As with the valepotriates and sesquiterpenes, the relative and absolute concentrations of these agents, as well as others contained in the root and extracts, vary among preparations.

## Pharmacokinetics

The chemicals thought to be most responsible for the central nervous system actions of valerian root are the valepotriates, particularly bornyl acetate, and the sesquiterpenes, most notably valerenic acid. As the valepotriates are generally lipophilic, these unstable compounds are found primarily in alcohol extracts, while the more hydrophilic sesquiterpenes are present in both aqueous and alcohol extracts. Both types of extracts are consumed for therapeutic purposes.

Given the variations in the chemical constituents among individual valerian plants, and the large variety of species, it is impossible to draw firm conclusions about the pharmacokinetic properties of valerian by studying extracts. Rather, emphasis is placed on examining the pharmacological characteristics of particular chemical constituents, principally bornyl acetate and valerenic acid.

Serum levels of valerenic acid peak one to two hours following oral administration of valerian to humans. Its elimination half-life is one to two hours, with trace amounts detectable in serum for up to five hours. These results confirm that valerenic acid is absorbed intact from the human gastrointestinal tract following consumption of a valerian extract, and indicate relatively rapid elimination from the bloodstream. Given these findings, it is recommended that for treating insomnia valerian be taken 30 minutes to two hours before retiring so that sleep is sought at a time when valerenic acid blood levels are rising or at their peak.[12]

Because insomnia is most commonly experienced in the elderly, valerenic acid levels have been quantified in older individuals after single and repeated administration of valerian extract. No significant differences in the peak serum concentration, time to peak serum concentration, and the elimination half-life were found for valerenic acid among these subjects following a single oral dose of 600 mg valerian. Differences in these indices were noted, however, among these subjects when the same dose was taken once daily for two weeks. This indicates that changes in the absorption and elimination of valerenic acid occur over time with continued administration. This might explain, in part, the variability in its effectiveness among individuals.[13]

While these data indicate that valerenic acid is detectable in blood following ingestion of valerian, they reveal nothing about whether this compound enters the brain. An in vitro study aimed at examining this question tested the extent of accumulation of three sesquiterpenes, valerenic, acetoxyvalerenic, and hydroxyvalerenic acids, in human brain tissue in comparison to diazepam, a benzodiazepine anxiolytic. The results revealed the sesquiterpenes are transported much more slowly across the blood-brain barrier than the benzodiazepine, presumably reflecting the fact that the valerian products are considerably less lipid soluble than diazepam. As it is known that diazepam readily penetrates into the brain by passive diffusion, it was concluded from this work that the sesquiterpenes probably gain entry by way of a specialized transport system.[14] While such work does not prove these compounds, or any other chemical constituents of valerian, appear in the human brain following oral administration, it does suggest these chemicals have only limited access to this organ. For this reason, it is critical that sufficient quantities of valerian be taken to achieve blood levels of valerenic acid necessary to ensure that enough accumulates in the brain to cause the desired effect. Because the therapeutically important valerian constituents, or combination of constituents, are unknown, and there is a high degree of variation in the chemical composition of different preparations, it is impossible to predict with any certainty the most consistently effective dose of valerian extract. Nonetheless, these pharmacokinetic studies indicate that oral administration of valerian results in the absorption of compounds that may penetrate into the brain and affect central nervous system activity.

Studies with rat tissue indicate that valerian valepotriates and sesquiterpenes, including valerenic acid, are extensively metabolized by the liver, and are substrates for drug transporters.[15-17] While data suggest that components of valerian extract interact with certain drug metabolizing enzymes, experiments indicate that, when consumed at the standard dose, valerian extract has little effect on the activities of these enzymes in humans.[18] Taken together, these findings suggest that the pharmacological response to valerian may be due to the formation of biologically active metabolites. Moreover, they suggest it is unlikely that valerian consumption will modify the response to most prescription and over-the-counter medications.

## Pharmacodynamics

The valepotriates and sesquiterpenes, and in particular valerenic acid, are thought to be responsible for the sedative and anxiolytic effects of valerian root extract. Evidence suggests these agents may induce these effects by influencing GABA receptor function in the brain. Upon release from central nervous system presynaptic neurons, GABA attaches to two different types of postsynaptic receptors. One, termed the $GABA_A$ receptor, is known to be the primary site of action of many central nervous system depressants, including the barbiturates, benzodiazepines, and ethyl alcohol (see Figure 7.1). The other, the $GABA_B$ receptor, while not directly influenced by these drugs, is stimulated by baclofen, an agent used to treat spasticity and that also depresses central nervous system function. Activation of either $GABA_A$ or $GABA_B$ receptors reduces the activity of the neurons on which they reside. Accordingly, compounds that enhance GABAergic activity are central nervous system depressants, whereas those that inhibit this receptor site are brain stimulants and convulsants.

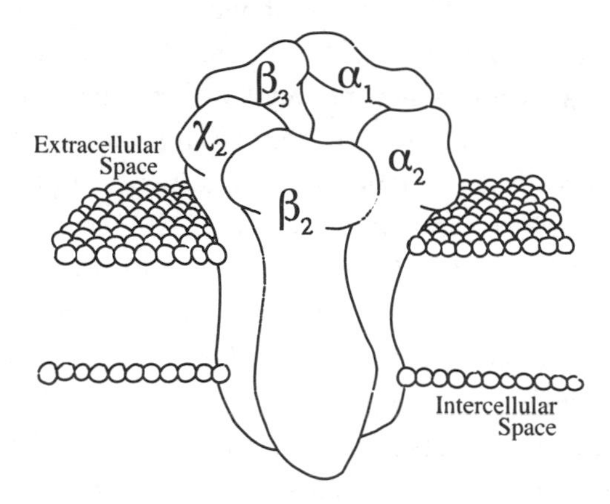


Figure 7.1 Representation of a $\beta_2$ / $\beta_3$ subunit-containing $GABA_A$ receptor, with indications as to the sites of action of various agents on this receptor subtype

The $GABA_A$ receptor is composed of five protein subunits (refer to Figure 7.1). As these are drawn from a pool of nearly two dozen distinct $GABA_A$ receptor proteins, a number of different subunit combinations are possible, with each one representing a molecularly and pharmacologically distinct $GABA_A$ site. The subunit pool utilized for the formation of $GABA_A$ receptors is composed of seven different classes based on their amino acid composition. Each of these classes is designated by a Greek letter. Thus, for example, there are six types of $\alpha$ subunits, and three types of $\beta$ and $\gamma$ subunits.[19] Because of differences in the subunit combinations of $GABA_A$ receptors, drugs acting at this site might affect one group of receptors but not another. This type of selectivity can account for differences in the pharmacological responses observed with agents that interact with various subtypes of the $GABA_A$ site. Thus, it appears that $GABA_A$ receptors containing an $\alpha_1$ subunit are most responsible for mediating the sedative and hypnotic responses to benzodiazepines, whereas anxiety is reduced with benzodiazepines that activate the $GABA_A$ receptors containing an $\alpha_2$ subunit.

A number of mechanisms of action have been proposed for valerian extract and its constituents. Suggested neurotransmitter sites of action include adenosine A(1), $serotonin_{5a}$, and $GABA_A$ receptors.[20-23] Of these, the $GABA_A$ receptor appears to be the most likely site of action. *In vitro* experiments demonstrate that valerenic acid, as well as valerian extracts, stimulate chloride currents in the synapse, a GABA-like action.[24-26] The effect of valerenic acid is selective for $GABA_A$ receptors containing $\beta_2$ and/or $\beta_3$ subunits (refer to Figure 7.1). The interaction of valerenic acid with $GABA_A$ receptor β subunits has been confirmed by a number of laboratories. This molecular target is a different site of action than that identified for the benzodiazepines, although the overall effect on $GABA_A$ receptor function is similar. The fact that a benzodiazepine site antagonist does not block the action of valerenic acid is proof that it does not interact with the same site as diazepam or other benzodiazepines. Also, this selective attraction between valerenic acid and the $\beta_2$ / $\beta_3$ subunit containing $GABA_A$ receptors indicates its pharmacology may differ somewhat from ethyl alcohol and barbiturates, as these compounds influence a much larger group of $GABA_A$ receptors.

In vivo laboratory animal studies reveal that valerian extract, and valerenic acid alone, decrease brainstem neuronal activity.[27] This finding is consistent with the notion that valerenic acid selectively enhances $GABA_A$ receptor function. Laboratory studies also support the proposal that the chemical components of valerian affect animal behavior. Thus, valerenic acid, as well as linarin, a flavonoid found in valerian root, decrease locomotor activity in mice and display sedative and hypnotic effects in rats and mice.[28-30]

The demonstration that valerian root extracts, as well as individual components of this mixture, display sedative and anxiolytic effects in laboratory animals is crucial for convincing investigators of its potential therapeutic utility. It also provides behavioral endpoints that can be used in searching for the mechanism of action of valerian. Various animal models were created to match behavioral modifications, usually in rats, with human responses. While there are significant limitations in extrapolating effects on animal behavior to humans, some models of anxiety are useful for predicting clinical responses. One such model is the plus maze test. The principle of this assay is that laboratory animals are uncomfortable in unfamiliar surroundings. This emotional response diminishes their normal tendency to explore their environment. When placed in the plus maze, the animals can freely move between an open and a closed space. The time it takes for them to voluntarily move from the closed, or protected, space to explore the open area is used as a measure of anxiety. Animals administered a known anxiolytic remain in the closed environment for a much shorter period than those not receiving a placebo. As for valerian, it has been reported that the extract itself, as well as pure valerenic acid, reduces the time rats spend in the closed arms of the maze, much like diazepam. Studies with genetically modified mice reveal that, unlike normal animals, those lacking the $\beta_3$ subunit no longer display an anxiolytic response to valerenic acid, confirming this receptor as the brain target for this compound, and the β subunit as the molecular site of action.[26] These results confirm these substances display anxiolytic potential.[30]

Clinical studies have been performed to examine the behavioral effects of valerian root extracts. Two meta-analyses of past human trials were undertaken to examine the effectiveness of the extract as a treatment for insomnia. One of these retrospective analyses reviewed 16 studies involving 1,093 patients. When taken together, the results indicated that valerian is of significant value in improving sleep.[31] The second meta-analysis involved a review of 18 randomized clinical trials. Again, the results suggested that valerian decreases the time it takes to fall asleep and modestly improves sleep quality.[32] As there was heterogeneity in the outcomes for different measures in these studies, it appears that some individuals are more susceptible to the hypnotic effect of valerian than others. Recently, for example, a clinical trial was conducted examining the possible hypnotic effect of valerian in 227 patients undergoing treatment for cancer. In this case, the extract failed to provide any improvement in sleep.[33] The extent to which the clinical condition of these subjects, their cancer therapy, the dose of valerian, or its source influenced this outcome is unknown. However, this negative finding illustrates that not all individuals may benefit from consumption of this herbal extract.

## Adverse Effects

Valerian root extract is generally considered a safe alternative to prescription anxiolytics and hypnotics. Side effects associated with valerian use tend to be comparable to those reported for a placebo, and are much fewer and less severe than the ones encountered with the benzodiazepines or barbiturates. Nonetheless, long-term use of valerian for insomnia is not recommended because psychological dependence can develop for any substance or activity that is perceived as essential for sleep. While no specific toxicities are associated with the continued administration of valerian, hospitalization can result if the product is taken in conjunction with an overdose of other medications.[34] The extent to which valerian consumption contributed to the toxic response to these drug combinations is unknown. Because valerian displays central nervous system depressant activity and appears to act at the same receptor as other members of this drug class, this extract should not be administered in conjunction with other central nervous system depressants, including alcohol. Such a combination could result in an additive effect on the brain, with, for example, the individual being more affected by the consumption of an alcoholic beverage than usual.

Although there is always the possibility that continued administration of any substance can influence drug metabolizing enzymes, and therefore the breakdown and response to other medications, this does not appear to be a major concern with valerian. While valerian has been reported to have a minimal effect on one type of drug metabolizing enzyme in humans, it appears to be without effect on another known to be particularly important in transforming drugs to less active metabolites.[35] Accordingly, based on published work to date, the consumption of valerian would not be anticipated to modify responses to over-the-counter or prescription drugs, except for central nervous system depressants. As discussed previously, the latter is the result of pharmacodynamic rather than a pharmacokinetic interaction.

## Pharmacological Perspective

Valerian has long been used for combating insomnia and alleviating anxiety. Clinical studies have confirmed these effects. As with all herbal supplements, the chemical composition of valerian root extract is determined by such factors as the plant species, the climate and soil in which it is grown, the time of harvest, the conditions under which the root is dried, and the method of extraction for isolating the volatile oil. Given these factors, and the fact that responses can vary among individuals to even purified substances, it is understandable that while some find valerian a useful anxiolytic and hypnotic, others do not. Indeed, in spite of the high likelihood of variation in the content of pharmacologically active constituents among valerian preparations, many responders are consistently satisfied with this product. Perhaps these individuals are particularly sensitive to the active components or, because of past experience with the product, have a lower threshold for a placebo response when unknowingly taking a less active form of the extract. The length of time valerian has been used for its central nervous system effects, the positive clinical results with this extract, and the laboratory studies indicating its chemical constituents interact with a brain neurotransmitter known to be associated with the action of central nervous system depressants, all suggest that it is a relatively safe and pharmacologically active plant extract. Additional studies are needed to identify and characterize further the chemical constituents in valerian root that can affect behavior and to standardize the extract mixture. This will not only make it possible to determine more accurately the most appropriate use and dose of this product, but also make more predictable the response to this supplement.

# 8

# Lemon Balm (Melissa officinalis)

Writing in the fourth century BC, Theophrastus described some of the characteristics of melissophyllon, the Greek term for lemon balm. As the literal translation of this term is "loved by bees," it illustrates that the ancients were, along with insects, attracted by the scent of this plant. This point was reinforced in the first century AD when Pliny, the Roman naturalist, advised in his *Natural History* that rubbing a hive with lemon balm leaves, or planting the herb near a hive, would keep bees from deserting the hive. In the eighteenth century the Swedish botanist, Carl Linnaeus, gave the name Melissa to the genus of plants that includes lemon balm.

A member of the mint family, lemon balm has for centuries been employed as an herbal remedy and for culinary purposes. Its popularity is not surprising as humans were naturally drawn to it as a potential food source because of its pleasant, citrus-like odor and flavor. These properties are due to its production of volatile chemicals, including citral, citronellal, linalyl acetate, and caryophyllene.

Lemon balm is purchased as the dried leaf, as a tea, oil, or extract. All of these products are preparations derived from the plant leaves. Lemon balm has been, and still is, used for a variety of purposes. These include as a flavoring for foods, including candies, desserts, and drinks. It is placed on the skin as a mosquito repellant and as a treatment for infections. Lemon balm is also taken orally, often in combination with other plant derivatives, such as alcohol or valerian, to reduce anxiety, facilitate sleep, and elevate mood. It is a popular constituent in aromatherapy, a form of alternative medicine. In aromatherapy, the vapors of essential oils, which are volatile substances obtained from plants, are inhaled or the oils are applied directly to the skin. The aim is to enhance mood and cognition, or to treat infectious disease. The extent to which such effects can be attributed to the lemon balm plant, as opposed to a placebo effect or a response to other products taken with it, requires further investigation.

## Botany

Lemon balm belongs to the mint family, Lamiaceae, previously known as Labiatae. The latter designation refers to the characteristic shape of the flower. Tubular at its base, the flower extends to form labia or lips. The current designation, Lamiaceae, is derived from the name of one genus, Lamium, a plant that like lemon balm is native to Europe and Western Asia.

Lemon balm is cultivated for its scent and the purported therapeutic properties of the leaves. The plant has been naturalized in the United States. It grows wild in uncultivated areas throughout North America. A perennial, the lemon balm plant sprouts from the roots in spring. The nearly white flowers are usually present from June through September. Like other herbs in the mint family, the essential oil is produced by small glands on the leaves. Both fresh and dried leaves are used for cooking and medicinal purposes, although drying reduces the content of the volatile substances. By midsummer, lemon balm plants are two to three feet in height. As with all members of the mint family, the lemon balm stalk is square, not round, and the leaves are borne oppositely rather than alternately.

## Therapeutic Uses

In the sixteenth century, John Gerard wrote that lemon balm tea drove away sorrows of the mind.[1] This reflects the historic use of this tea for its calming and sedative properties. Even today, a bedtime cup of lemon balm tea is a popular treatment for insomnia. While lemon balm has for centuries been used to facilitate sleep, few placebo-controlled clinical trials have been performed to establish its effectiveness in this regard.[2]

The anti-anxiety effect of lemon balm extract has been the subject of clinical investigations, with particular emphasis placed on its ability to treat the agitation associated with dementia in general, and Alzheimer's disease in particular.[3] Over the centuries, lemon balm tea has also been used as an antipyretic to reduce fevers associated with infectious diseases, and as an anti-inflammatory agent and analgesic for treating gout and other inflammatory conditions.

## Constituents

Lemon balm leaf produces dozens of chemicals, at least 70 of which have been identified. Several of these are thought to be responsible for the therapeutic effects attributed to this plant. The compounds found in the lemon balm plant are divided into four major chemical classes: monoterpenes, triterpenes, phenylpropanoids, and flavonoids. The essential oil from lemon balm contains a large number of monoterpenes and triterpenes. Because monoterpenes are volatile they contribute to the distinctive odor of the plant. It is the monoterpene citral that gives the plant its lemon-like scent. Citral is a mixture of two stereoisomers, citral A, or geranial, and citral B, or neral. Citronellal is another monoterpene produced in the lemon balm leaf. Oleanolic and ursolic acids are lemon balm triterpenes of possible pharmacological interest. Rosmarinic acid is a lemon balm phenylpropanoid that is thought to have therapeutic activity. It is found in the leaves of several genera of the mint family, including rosemary, from which the name rosmarinic acid is derived. The flavonoids, some of which are believed to possess anti-inflammatory properties and to neutralize free radicals, are found in many plant species.

The concentrations of these ingredients in an herbal preparation of lemon balm can vary two to threefold, depending on the age of the plant at harvest and whether these constituents were derived from the entire plant or just the leaves.[4] The relative quantities of the lemon balm chemical components present in, for example, the essential oil, also vary with the method of extraction.[5]

Citral A and B are the most common monoterpenes in the essential oil of lemon balm, representing more than 50% of all constituents in this product. Citronellal and the sesquiterpene beta-carophyllene each represent approximately 6% of the chemicals in the oil.[6]

Rosmarinic acid comprises about 1.5 % of the dry leaf mass.[7] It, along with ursolic and oleanolic acids, are some of the major constituents of lemon balm leaf and leaf extracts.[8] Not all of the scores of chemical constituents contained in the lemon balm leaf extract have yet been identified or pharmacologically characterized.

## Pharmacokinetics

Studies have been conducted on the absorption and metabolism of some of the chemicals in lemon balm products. It was found in rats that citral is completely absorbed following oral administration, but that it is rapidly metabolized.[9] Within five minutes of an intravenous injection in rats virtually the entire dose of citral is transformed into metabolites. The compound and its metabolites are excreted primarily in the urine and bile.[9] These findings suggest that citral is unlikely to play any significant role in a therapeutic response to lemon balm.

When administered to rats orally, ursolic acid was rapidly absorbed and displayed a half-life in blood of between one and four hours.[10,11] Widely distributed throughout the body, the highest concentrations of ursolic acid were found in the lung, spleen, and liver. A significant amount was also detected in rat brain, indicating this compound is capable of crossing the blood-brain barrier. While these data suggest that ursolic acid might affect brain function, its rapid elimination from the blood suggests that any pharmacological response would be brief.

Following oral administration of the plant extract, or its consumption as the purified compound, rosmarinic acid was found to be poorly absorbed and rapidly eliminated in rats.[12,13] Data indicate that rosmarinic acid is metabolized by intestinal organisms so that only a small portion of an orally administered dose is available for passive diffusion into the systemic circulation.[14] The metabolites formed in the intestines, which include *meta*-coumaric and hydroxylated phenylpropionic acids, are transferred to the blood by an active transport system. It is unclear whether these metabolites display any pharmacological activities on their own. In rats, the majority of the rosmarinic acid absorbed into the blood is extensively metabolized in the body and excreted in the urine.[15,16] Taken together, these findings suggest that rosmarinic acid is an unlikely candidate as the constituent of lemon balm primarily responsible for any effects on the central nervous system.

The half-life of oleanolic acid following intravenous administration to various animal species is relatively short, ranging from 12 minutes in rats to 57 minutes in dogs.[17] From these data it was predicted that the half-life in humans would be somewhere between 45 minutes and three hours. In a study involving 18 healthy volunteers, the measured half-life was up to eight hours following oral administration. However, this value was highly variable, with some subjects displaying much shorter half-lives.[18] Nonetheless, these results indicate that oleanolic acid may remain in the body of some individuals for a sufficient time to induce a pharmacological effect. The wide variation in half-lives in humans suggests a behavioral response to oleanolic acid would be unpredictable.

In vivo and in vitro laboratory animal studies indicate that lemon balm constituents can affect the absorption, distribution, and metabolism of other chemical agents, such as prescription medications.[19-22] Citronellal, for example, inhibits intestinal transporters responsible for the absorption of digoxin, a drug used to treat heart failure. Likewise, ursolic acid interacts with transporters involved in eliminating rosuvastatin, a cholesterol-lowering agent, from blood. Citral has been shown to be an inhibitor of a liver enzyme responsible for the metabolism of a variety of drugs. Lemon balm extract may also inhibit the absorption of thyroxine, a thyroid hormone. These data suggest that consumption of lemon balm can influence the responses to certain drugs, either prolonging or enhancing their actions, or reducing their effectiveness. However, given what is known about the pharmacokinetics of some of the major lemon balm constituents, it would probably be necessary to consume large quantities of the herb over a prolonged period of time to achieve the sustained blood levels needed to influence the response to conventional medications. The fact that there are few reports of adverse drug interactions associated with lemon balm consumption reinforces the conclusions drawn from pharmacokinetic studies. That is, the known constituents of this plant appear to be poorly absorbed, extensively metabolized, and rapidly cleared from the body, making them unlikely candidates as drugs or toxic agents.

## Pharmacodynamics

A number of laboratory animal studies have been undertaken to define the responses to, and possible mechanisms of action of, lemon balm extract and its chemical constituents. As for pharmacological effects, oral administration of either an ethanol extract of lemon balm or purified rosmarinic acid reportedly induces analgesia in mice.[23] In rats, however, others have found rosmarinic acid to be ineffective as an analgesic and anti-inflammatory agent.[24] In contrast, a significant anti-anxiety effect was detected in rats receiving an intraperitoneal injection of rosmarinic acid.[25] Citronellal and citral have been reported to induce sedative and analgesic effects in mice, and analgesic and anti-inflammatory effects in rats.[26,27] The conflicting results from these and other animal studies and the historic use of lemon balm by humans, suggest that consumption of this extract at the proper dose may modify central nervous system functions.

With regard to mechanism of action, an in vitro study revealed that an aqueous extract of lemon balm inhibited GABA transaminase in rat brain tissue.[28,29] As GABA transaminase is the enzyme responsible for the metabolism of GABA, an inhibitory neurotransmitter, its blockade results in the elevation of brain GABA and a decrease in central nervous system activity. It has been suggested that rosmarinic, ursolic, and oleanolic acids are the lemon balm constituents responsible for inhibiting GABA transaminase.[7] It has also been reported that chronic administration of lemon balm extract to mice causes a decrease in serum corticosterone, a hormone associated with stress, suggesting a possible relationship between this action and its anti-anxiety effects.[29] If inhibition of GABA transaminase occurs in vivo, such an action would provide a plausible explanation for the anti-anxiety and sedative effects reported for lemon balm. Proof of this hypothesis awaits definitive data demonstrating that inhibition of this enzyme occurs in animal brain following systemic administration of lemon balm or its constituents, and that the herbal extract prevents seizures, an established response following blockade of GABA transaminase. Until then, these findings must be considered tentative as an explanation for the mechanism of action of this product.

Another proposed mechanism for the central nervous system effects of lemon balm is an interaction with the brain acetylcholine neurotransmitter system. In vitro work demonstrated that rosmarinic acid is capable of inhibiting acetylcholinesterase, the enzyme responsible for the destruction of the neurotransmitter acetylcholine.[30] Such an action would prolong the action of this neurotransmitter, which would explain the reported beneficial effects of lemon balm on the cognitive impairments associated with Alzheimer's disease. However, other investigators have been unable to demonstrate that lemon balm is capable of inhibiting acetylcholinesterase in vitro, although they did note that it directly interacts with muscarinic and nicotinic acid binding sites in human brain tissue, both of which are acetylcholine receptors.[31] These investigators propose that lemon balm constituents activate acetylcholine receptors, an action that could be responsible for enhancing memory. While both the acetylcholinesterase inhibition and acetylcholine receptor activation hypotheses are intriguing, neither effect has been demonstrated in vivo following systemic administration of conventional doses of lemon balm extract. Additionally, unless these actions are restricted to the brain, for which there is no evidence, a generalized activation of the acetylcholine system throughout the body would be expected to result in a number of discomforting side effects, including changes in blood pressure, lacrimation, or tearing, salivation, or drooling, and increases in urination and defecation. Because none of these symptoms is typically associated with consumption of lemon balm, it seems unlikely that constituents of this extract are generalized activators of the acetylcholine system at the doses employed.

In attempting to establish a mechanism of action for lemon balm as a treatment for Alzheimer's disease, rosmarinic acid was fed to mice that had been genetically modified to produce excessive amounts of beta-amyloid in the brain. Beta-amyloid is a protein known to be present in abnormally high quantities in the brains of individuals with Alzheimer's disease. Consumption of rosmarinic acid for ten months decreased brain accumulation of beta-amyloid in these mice as compared with untreated control subjects.[32] The clinical significance of this discovery is uncertain, however, as there are questions about the relevance of this mouse model to the human condition, and the causal relationship between the accumulation of beta-amyloid and Alzheimer's disease is a matter of controversy.

Other laboratory studies have been conducted to define the antioxidant, antimicrobial, analgesic, anti-inflammatory, and anticancer actions of lemon balm.[26,27,33-39] It is known that oxidative stress and the generation of free radicals are associated with a number of disease states, including type 2 diabetes and central nervous system disorders. This has led to an interest in identifying antioxidants, which are compounds capable of attenuating these biochemical responses. Such agents would be expected to halt or slow the tissue damage associated with trauma, disease, and neurodegenerative disorders. The results of in vitro experiments suggest that methanol extracts of lemon balm are more effective than aqueous extracts in scavenging free radicals and blocking lipid peroxidation, a process that generates free radicals. Reports indicate that the monoterpenes citral and citronellal, besides displaying antimicrobial activity, appear to be the most potent inhibitors of lipid peroxidation in lemon balm extract. Administration of the essential oil to diabetic mice reduced blood sugar and increased serum insulin levels, suggesting it may be of some benefit in controlling diabetes.

Because the extent to which lemon balm extract displays antioxidant activity in vivo is unknown, it remains questionable whether its consumption provides any protective benefit from the ravages of free radicals. Likewise, definitive in vivo human data are lacking to demonstrate that the antimicrobial, anti-inflammatory, analgesic, anti-diabetic, and anticancer effects detected in laboratory animals are of any clinical significance.

As for human studies, the purported calming effect of lemon balm tea provided the rationale for examining its effects on the anxiety and agitation associated with neurological disorders. A double-blind, placebo-controlled trial was performed to assess the value of lemon balm aromatherapy in 70 patients with agitation associated with severe dementia and Alzheimer's disease. A significant improvement in the quality of life was noted for all 35 subjects who underwent lemon balm aromatherapy as compared to only 11 of the control patients who received aromatherapy with odorless sunflower oil.[40] Because of its safety profile, it was suggested that lemon balm may be useful for the treatment of dementia, either by itself or in conjunction with other medications.[41] This positive effect was not replicated by the same investigators in a later study of 114 agitated patients.[42] In this case, the quality of life was improved whether the patients were exposed to lemon balm or placebo aromatherapy. It was concluded that the apparent benefits observed in both studies were due primarily to the positive effects of the human interactions with these subjects in association with the clinical trials, rather than to the aromatherapy. While it is possible individuals may perceive some benefit from lemon balm aromatherapy, the lack of data indicating the appearance of lemon balm constituents in blood following inhalation of the essential oil vapors makes it difficult to attribute any positive response to a pharmacological effect of the herb.

Clinical studies on mood and memory have been conducted following oral administration of lemon balm to healthy subjects. In a placebo-controlled study, 20 volunteers received a single dose of placebo or 600, 1000, or 1600 mg of dried lemon balm leaf. Enhanced calmness and an increase in memory were reported one, three, and six hours following consumption of the highest dose of dried lemon balm leaf.[31] It has also been reported that combined treatment with lemon balm and valerian, an herb discussed in Chapter 7 of this volume, significantly reduced anxiety in healthy subjects at lower, but increased anxiety at higher, doses.[43] One interpretation of these findings is that because exposure to a pleasant odor or taste, regardless of the source, has a calming effect on normal subjects, there is no need to attribute these findings to a specific pharmacological effect. The results with the valerian combination treatment are particularly difficult to interpret inasmuch as valerian is known to have anti-anxiety effects on its own, making it impossible to draw any firm conclusions about the contribution of lemon balm to the observed effects. Thus, it could be that the purported beneficial effects of lemon balm are more related to its agreeable odor and taste than to a direct action of plant constituents on the brain. That is, the perceived benefits may be more related to a psychological, rather than to a pharmacological, effect.

## Adverse Effects

No significant adverse effects are associated with the human consumption of lemon balm. In vitro and in vivo laboratory animal studies indicate that chemicals contained in lemon balm extract and essential oil have the potential to modify the absorption, metabolism, and elimination of other agents, such as prescription drugs. However, the pharmacokinetics of the lemon balm constituents are such that they would not be anticipated to be present in sufficient concentrations for a sufficient period of time to modify drug responses in humans. This explains why there are few reports of adverse interactions between lemon balm and conventional medications.

## Pharmacological Perspective

It is not surprising that a plant such as lemon balm, which produces a variety of chemicals and has a pleasant odor and taste, has been reported to display a number of positive pharmacological effects, ranging from sedation to analgesia. As documented for centuries, many individuals find that consumption of this herb has a calming effect and facilitates sleep. However, it is notable that the constituents of lemon balm thought to be responsible for these responses are found in a number of other plants. Given their widespread distribution throughout the plant kingdom, it seems likely that any definitive and consistent clinical benefit associated with one or more of these compounds would be obvious, easy to verify clinically, and found in association with many plant products. This is not the case. Furthermore, there is no precedent for a single compound, or group of chemicals, displaying such a wide variety of beneficial actions such as those attributed to lemon balm without noticeable side effects. Any significant effect on brain neurotransmitter systems, inflammation, or pain pathways is bound to have undesirable consequences in at least some individuals at certain doses. Thus, the fact that lemon balm appears innocuous can be taken as evidence that there are few, if any, clinically active substances in this plant extract or the essential oil. The pharmacokinetic data suggest that even if certain lemon balm constituents are pharmacologically active, they are poorly absorbed into the systemic circulation following oral administration and rapidly eliminated. These properties lessen the likelihood that they will achieve and sustain sufficient concentrations at the appropriate target sites in the human brain. While there are in vivo animal data indicating useful pharmacological responses to lemon balm, such findings are not always applicable to humans because of pharmacokinetic and pharmacodynamic differences among species and the difficulties of replicating human disorders in laboratory subjects. Moreover, it is unclear whether the doses of this herbal supplement shown to be effective in animal studies are equivalent to those taken by humans. Of course, it remains possible that some as yet unidentified constituent of lemon balm, or a metabolite of one of the chemicals produced by this plant, is responsible for its actions. However, given the available data, it seems likely that the reported calming and hypnotic actions of lemon balm are nonspecific effects related to its pleasant odor and taste, or to the power of suggestion, rather than to a specific, predictable, and consistent pharmacological response.

# 9

## Kava (Piper methysticum)

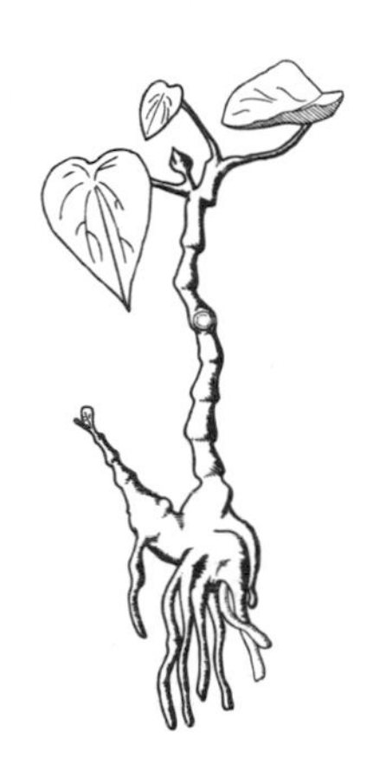

In the autumn of 1835, Charles Darwin traveled from the Galapagos Islands to Tahiti during his five year voyage on the *HMS Beagle*. He describes in his journal a walking tour of Tahiti with local guides. Besides viewing stands of wild sugar cane, Darwin notes seeing kava growing at the edge of a small creek. He recognized this plant from a description provided by Johann Forester, the botanist who accompanied Captain James Cook on the *HMS Endeavour* on an earlier expedition to the South Pacific. Forester named the plant Piper methysticum. The species designation was derived in part from the Greek word *methi*, meaning strong drink. Knowing this, Darwin tasted the plant stem. He described the flavor as acrid and unpleasant.[1] However, it is the root of the kava that is typically consumed for its central nervous system effects. These purportedly include a relief from anxiety, a feeling of tranquility, and somnolence. Usually the root is macerated or ground and an aqueous extract served as a drink. Alternatively, the unprepared root is chewed.

A native of tropical islands, the precise geographical origin of kava, the discovery of its psychoactive properties, and the development of its use are unknown as written records on these subjects date only from the eighteenth century when the British first arrived in Oceania. As its ingestion causes noticeable and generally pleasant changes in the sensorium, it was undoubtedly used by South Sea island natives for centuries before the arrival of Europeans. Evidence for this is that both Captain Cook and Darwin found the locals consuming kava as part of well-established social and religious ceremonies. Today kava beverages are still used for social events in Tonga, Samoa, and Fiji. While once generally available in the Americas and Europe as an herbal supplement, the use of kava is now restricted because of reports linking it to severe liver toxicity. In fact, safety concerns led to its ban in several countries, including France and Switzerland.

As kava products are marketed as anxiolytics and hypnotics, the active constituents, thought to be kavalactones, appear to be central nervous system depressants akin to ethyl alcohol, the barbiturates, and the benzodiazepines. The kavalactones of particular interest are desmethoxyyangonin and kavain because of the evidence they may influence neurotransmitter activity in the brain. It is unknown whether the central nervous system effects of kava are solely attributable to kavalactones or to other plant constituents as well.

## Botany

The kava plant, Piper methysticum, belongs to the pepper family, Piperaceae. The best known member of the Piperaceae is Piper nigrum, the seeds of which are the source of common black pepper. Several of the approximately 2,000 plant species in the genus Piper are known to affect central nervous system function. All members of this group grow in humid, subtropical regions. Piper nigrum was originally found in southern India and is cultivated on the Malay peninsula and the West Indies. A native of Oceania, Piper methysticum is a cultivated plant known variously as kava, kawa, or ava, depending on the local dialect. Kava is probably a hybrid or mutant of Piper wickmanni, a related wild species native to Vanuatu, an island that lies northeast of Australia and west of the Fiji Islands. Because its seeds are sterile, kava reproduces vegetatively and is cultivated by cuttings.

Its present geographic distribution is presumably the result of its being transported and transplanted by humans.[2]

A perennial, kava is a dense, shrublike plant that grows to 10 to 15 feet in height. The leaves are large and relatively sparse. Male and female flowers are rarely produced, and the seeds are infertile. Various cultivars have been selected from naturally occurring stocks. The cultivars differ in the concentrations of constituents thought to mediate the central nervous system effects of kava. Moreover, the kavalactone content is known to increase with the age of the plant. The kavalactones are produced primarily in the roots, rootlets, and the stalk just adjacent to the root system. Typically, the root is harvested for preparing a kava extract when the plant is three to four years of age. The potency and activity of a kava product varies significantly among batches and producers as a function of the cultivar that it harvested, its age, and the portion of the plant used to prepare the extract.

## Therapeutic Uses

Just as ethyl alcohol is used to increase sociability, so an aqueous preparation of kava root was, and is, consumed socially by some south Pacific islanders. For generations, and more likely thousands of years, Oceanic cultures also included a kava beverage as a component of ceremonial rituals associated with, for example, births or deaths. Kava drinking was popular in Europe and the United States throughout the twentieth century, first because of its reported benefits as a treatment for insomnia and anxiety, and later as part of the growing interest in herbal supplements in general. While today the product is taken primarily for its effects on central nervous system function, there are historical reports suggesting that kava leaves and roots were once used to make poultices for wound healing and for treating headaches.[3] Concerns about a possible association between kava use and serious, and sometimes fatal, liver toxicity have markedly reduced enthusiasm for this product.

Clinical studies indicate that kava extract facilitates sleep. As compared with what occurs with many prescription hypnotics, a kava-induced sleep appears less likely to be associated with a morning hangover.[4,5] Studies have also suggested that kava extract may be effective for the management of anxiety.[6]

## Constituents

The kava plant, like many other species of Piper, produces a number of pharmacologically active ingredients. The central nervous system effects of kava are generally attributed to six of the eighteen kavalactones found in this plant. These are kavain (see Figure 9.1), dihydrokavain, methysticin, dihydromethysticin, yangonin, and desmethoxyyangonin. The many different cultivars of kava grown throughout the South Pacific produce widely differing amounts of the kavalactones.

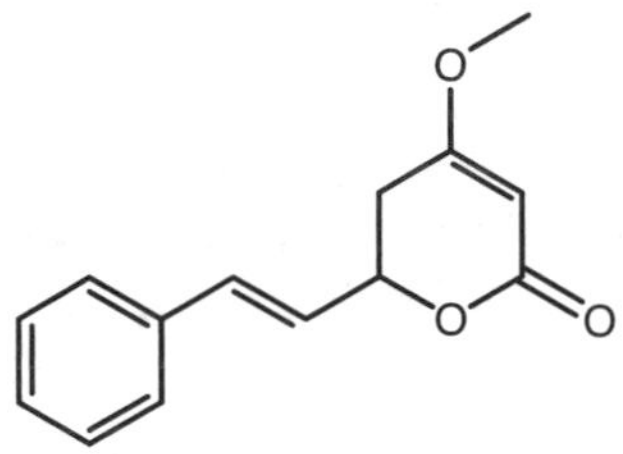

Figure 9.1 Chemical structure of kavain (Wikipedia)

These six kavalactones represent 9% to 12% of the weight of the dried rootstock. In most samples dihydrokavain is present in the highest concentration, representing approximately 30% of the total kavalactone content. It is followed by kavain and methysticin at 20% each, dihydromethysticin at 15%, yangonin at 10%, and desmethoxyyangonin at 5%. Thus, dihydrokavain, kavain, and methysticin comprise the majority of kavalactones in the dried root of the kava plant.[7]

These six kavalactones vary with regard to their activity on the nervous system function. In a study of 121 cultivars from 51 Pacific islands, those most preferred for their psychoactive effects were the ones having the higher content of kavain, dihydrokavain, and methysticin.[8]

## Pharmacokinetics

The kavalactones differ from one another with respect to the extent of their absorption following oral administration and in their duration and sites of action within the central nervous system. The absorption of the kavalactones in humans following oral administration appears to be more complete when they are consumed together in the plant extract as compared to when they are taken individually. Whether this is because one or more of the kavalactones facilitates the transfer of the others across the intestinal lumen or the extract contains an as yet unidentified constituent that enhances absorption remains to be determined. It is possible that a constituent in the extract slows its passage through the gastrointestinal tract, increasing the amount of time available for absorption. An orally administered 200 mg dose of kavain taken in a kava extract has an elimination half-life of nine hours in humans.[9] This suggests that kavain is present in the body for a sufficient amount of time to exert a pharmacologically meaningful response, assuming the amount consumed yields the brain levels necessary to modify central nervous system activity. Very little of an orally administered dose of kavalactones is metabolized in humans, with the majority excreted unchanged in urine.[7]

Laboratory animal experiments indicate that while individual kavalactones are transported across the gastrointestinal tract following oral administration, the amount of the total dose that appears in blood is quite low. As with humans, absorption is greater if the kavalactones are administered as a mixture rather than individually. For example, when kavain alone is administered by mouth to rats at a dose of 100 mg/kg, the maximum blood level is 2.6 microgram/ml, whereas the blood level is double that when the same dose of kavain is given orally in a kava extract. Because repeated administration of kavain does not result in its accumulation in the body, it is not stored in tissue. This accords with its being a relatively lipid insoluble compound.[10]

These findings, along with those reported for humans, suggest that consumption of a kava extract yields higher blood, and presumably brain, levels of kavalactones than administration of an equivalent dose of an individual kavalactone. Because the reason for this is unknown, it remains possible that an extract constituent is responsible for this enhanced absorption of kavalactones. However, such a component may also increase the absorption of toxins that may be present in the plant.

As the limited absorption and storage of kavalactones suggest low lipid solubility, it could be assumed these agents may have difficulty in crossing the blood-brain barrier and in penetrating into the brain. Experiments in mice, however, demonstrate directly that the intraperitoneal injection of kavain, dihydrokavain, desmethoxyyangonin, or yangonin yields within 5 minutes measurable brain levels of each. In this experiment the brain content of dihydrokavain was highest, and the brain level of yangonin the lowest. As the compounds that penetrate most readily into the brain are more rapidly eliminated, it is likely these agents are gaining entry by passive diffusion, with the more lipid soluble agents entering and exiting more easily than those that are more water soluble.

Administration of the kavalactones in a kava resin mixture increases the brain concentrations of kavain and yangonin, but not of dihydrokavain and desmethoxyyangonin.[11] The reason for this differential effect with the resin preparation is unknown. In light of these brain penetration studies it is noteworthy that kavain and dihydrokavain are generally more psychoactive than yangonin and desmethoxyyangonin. While not definitive proof, this correlation between the ability to penetrate into brain and central nervous system activity suggests that a sufficient amount of these substances, or some other active agent, gains access into the brain following systemic administration of a kava extract. While these brain penetration studies indicate the kavalactones gain entry into the central nervous system, they do not prove that the concentrations attained in human brain at the doses normally taken are sufficient to account for their reported anxiolytic and sedative effects.

## Pharmacodynamics

Laboratory animal studies have been performed to define the pharmacological properties and mechanism of action of the kavalactones. It has been reported that methysticin and dihydromethysticin are anticonvulsants and sedatives. It has also been shown that orally administered kava extract minimizes the degree of brain damage in rodents that results from an experimentally induced ischemic episode.[12] Further evidence that administration of kava extract can affect brain function is provided by the finding that its injection alters dopamine levels in select areas of the brain, in particular the nucleus accumbens, a limbic system region.[13] The possible involvement of the nucleus accumbens in the response to kava is interesting in that this brain area is known to play an important role in mediating feelings of pleasure and mirth, as well as aggression and fear. The nucleus accumbens also contributes significantly to the placebo response to inert substances. Thus, the responses to kava extract could very well be due to a direct or indirect effect on this brain region.

Several mechanisms of action have been proposed for the sedative and anxiolytic effects of kava extracts. There have been studies on the possible interaction of kavalactones with the GABA neurotransmitter system in the brain as this is the site of action of many central nervous system depressants, including ethyl alcohol, the benzodiazepines, such as Librium® and Valium®, and the barbiturates. In one study it was found that kavalactones, like the benzodiazepines, enhance the attachment of GABA to its receptor in the brain.[14] This explanation of its mechanism is doubtful because the concentrations of kavalactones necessary to induce this biochemical effect are more than 1,000 times greater than for the benzodiazepines. Given the limited ability of kavalactones to penetrate into the brain, it seems unlikely that such concentrations would be attained at the doses normally administered. Also, the reportedly large separation between the anxiolytic and sedative doses of the kavalactones tend to belie a generalized effect on GABA receptors. Nonetheless, these findings leave open the possibility that the kavalactones may selectively activate a subtype of the GABA receptors in the brain at much lower concentrations. Such an action could explain their pharmacological responses. However, this remains only a theoretical possibility until direct proof of a selective GABA receptor interaction is obtained.

There are also data indicating that the kavalactones influence voltage-gated ion channels in the brain.[15,16] Reports suggest that these compounds are capable of blocking neuronal calcium, potassium, and sodium channels, all of which play an important role in regulating neuronal excitability. It is speculated that such an action could indirectly influence GABA activity and might be responsible for the anticonvulsant and anxiolytic effects of the extract. As with the GABA receptor, these effects on ion channel activity are observed only at relatively high concentrations of kavalactones, calling into question their relevance with respect to their effects in humans. Also, as was the case with the GABA receptor results, a generalized inhibition of voltage-gated ion channels would have catastrophic effects on brain function.

Given the relative safety of the kavalactones, at least with regard to central nervous system activity, a nonselective inhibition of these ion channels would seem to be an unlikely mechanism of action. It remains possible, however, that the kavalactones may, at lower concentrations, specifically interact with one or more of the voltage-dependent ion channels in the brain, dampening neuronal firing in a selective manner.

Kavalactones have also been found to influence the brain levels and actions of some neurotransmitters, including norepinephrine, serotonin, and dopamine. It has been shown that kavain is capable of inhibiting the neuronal accumulation of serotonin and norepinephrine, although the concentrations needed are quite high relative to those required for drugs such as fluoxetine (Prozac®) and desipramine (Norpramin®), antidepressants known to have such an effect on one or both of these transmitter systems.[17] This makes it appear unlikely that inhibition of neurotransmitter reuptake is the mechanism of action of kavalactones. In vitro, the kavalactones reportedly inhibit the neurotransmitter metabolizing enzyme monoamine oxidase (MAO).[18] While such an action could contribute to their pharmacological effects, drugs known to be MAO inhibitors do not display the same pharmacological actions as the kavalactones. This might be because the drugs have other properties that are responsible for their distinct clinical characteristics, or that inhibition of MAO is unimportant with regard to the actions of kavalactones. The relevance of this action with respect to kava extract would be more convincing if MAO inhibition was shown to occur in vivo following administration of kavalactones at doses known to influence behavior.

Another reported effect of kavalactones is inhibition of NF-kappaB, an important component of the immune system.[19] This action is thought to explain the purported ability of kavalactones to protect against the development of cancer and to minimize brain infarctions in experimental models of stroke. An interaction with the NF-kappaB system is unlikely to contribute to the acute anxiolytic and hypnotic effects of kava.

Kava extract has been employed mostly as an anxiolytic and is favored over prescription drugs for this purpose because of a lower incidence of side effects, particularly sedation, at the doses employed to relieve anxiety. The results of clinical studies suggest that kava extract is effective in calming patients who suffer from an anxiety disorder or insomnia.[20-24] In one study a kava extract compared favorably with buspirone, an anxiolytic, and opipramol, an antidepressant and anxiolytic, in reducing anxiety in a group of 129 individuals with generalized anxiety disorder.[20] Likewise, when a kava extract was administered for three weeks it reduced anxiety and depression in a group of 60 patients with generalized anxiety disorder.[21] In a separate study, kava extract was reported to have significant beneficial effects in anxious patients when administered over a four week period.[22] A multicenter, double-blind clinical trial compared daily administration of placebo with kava extract in 61 patients diagnosed as having a sleep disturbance. The extract treatment was superior to the placebo over the four week treatment period, both in terms of improving the quality of sleep and in reducing anxiety.[23] A retrospective analysis of six placebo-controlled trials of kava extract concluded it is an effective treatment for nonpsychotic anxiety disorders.[24]

Not all clinical studies with kava have yielded positive findings. A review of six clinical trials indicates that kava extract was effective as a treatment for generalized anxiety disorder in four, but of no benefit in two.[25] A separate review of 24 studies involving patients suffering from an anxiety disorder or depression and in some cases healthy volunteers exposed to anxiety provoking situations, found that kava extract was effective in more than two-thirds of the randomized control trials.[6] In some cases similar positive responses were obtained with the kava extract and placebo, highlighting the difficulties associated with conclusively demonstrating the clinical effectiveness of any product purported to display anxiolytic activity. The lack of effectiveness in some cases could be explained by variations in the underlying pathologies or to differences in the content of active constituents in the kava extracts employed. Overall, the weight of clinical evidence suggests that consumption of kava extract may be of benefit for some individuals with anxiety or sleep disorders, and supports the long-held notion that this plant product has central nervous system depressant properties similar to alcohol, prescription anxiolytics, and hypnotics.

## Adverse Effects

That kava extract has been used as a beverage for possibly thousands of years suggests it is safe for consumption. This conclusion is reinforced by the many clinical reports indicating few, if any, side effects associated with its use as a treatment for anxiety and insomnia. Given this history, many were surprised with reports suggesting that severe liver toxicity may be associated with its use. This finding was particularly alarming because the toxicity is acute and irreversible, making it impossible to predict its onset or to treat it once it has occurred. The condition is characterized by an inflammatory response associated with a massive infiltration of lymphocytes, eosinophils, and macrophages resulting in liver necrosis, or cell death.[26] These reports relate to toxicity experienced by approximately 50 individuals, mostly Europeans, who were known to have consumed the extract. As a result, the United States Food and Drug Administration issued a warning about the use of kava, and some European countries had it withdrawn from their markets.

Several mechanisms for this hepatotoxicity have been proposed. These include the possible formation of kavalactone toxic metabolites or the presence in root extracts of some other toxic agent, such as pipermethystine, an alkaloid, or flavokawain B, a chalcone. Other possibilities are mold contaminants, in particular aflatoxins, which could appear during storage.[27]

The data supporting pipermethystine as the responsible toxin are unconvincing. This compound is one of three piperidine alkaloids isolated from aerial parts of kava. The other two are 3-alpha, 4-alpha-epoxy-5 beta pipermethystine, and awaine. Pipermethystine is concentrated in stems and leaves, not in the root.[28] In a study of the presence of pipermethystine in kava root extracts, none was detectable. Leaf extracts of Piper methysticum, on the other hand, contain 0.2% pipermethystine. Assuming the product that caused the liver toxicity was solely a root extract, it was concluded that the effect could not be due to pipermethystine.[29] This conclusion was bolstered by the finding that administration of pipermethystine to rats for two weeks had no significant effect on liver function, although there were signs of oxidative stress in some other organs.[30] While it is possible that pipermethystine is more toxic in humans than rats, the weight of evidence indicates the quantities of this toxin present in conventional kava extract preparations are too low to cause irreversible liver inflammation.

Studies suggest that flavokawain B is a more likely suspect as the offending toxin. It was found that the concentration of flavokawain B is significantly higher in an ethyl alcohol kava root extract than in an aqueous extract. Moreover, it has been shown that flavokawain B induces liver toxicity in mice by inhibiting the expression of NF-kappaB and by inducing a deficiency of glutathione, an endogenous antioxidant in liver.[31] An effect on glutathione levels has also been noted in response to in vitro exposure to kava extracts alone, and in combination with acetaminophen, an analgesic and antipyretic known to cause acute liver failure in humans.[32,33] It is unclear whether these results are due to the presence of flavokawain B in the test extracts, or whether other components of the extract, or combinations of components, cause the liver damage in vitro. Nonetheless, these data suggest that one possible explanation for the liver toxicity is that the samples consumed were ethyl alcohol rather than aqueous extracts of kava, and therefore contained quantities of flavokawain B sufficient to cause toxicity in susceptible individuals. It is also possible the samples were extracts of not only the kava root, but also the leaves and stems as well, which would have added pipermethystine to the sample. This would explain why liver toxicity is seldom, if ever, reported in association with kava use in the South Pacific because only aqueous extracts of the root are used.[26] These findings highlight the importance of purchasing herbal supplements from reputable dealers who provide basic information on the chemical composition of the product and the source of the sample.

There may also be adverse consequences from the consumption of kava extracts because some of the constituents interact with endogenous drug metabolizing enzymes. This can lead to a change in the levels of kavalactones or of prescription medications, increasing the likelihood of kava side effects or complicating the management of other conditions. It is known, for example, that some individuals metabolize kavalactones poorly because of a genetic deficiency in a particular drug metabolizing enzyme. Somewhere between 12% to 21% of all Caucasians have this deficiency, compared with less than 1% of the Asian/Pacific islander population.[26] Thus, on average, Caucasians are less able to metabolize kavalactones than Asians. The extent to which this genetic difference affects the response to kava extract, and the likelihood for side effects, is unknown.

It has been established that desmethoxyyangonin and dihydromethysticin increase the levels of certain endogenous drug metabolizing enzymes.[34,35] For this reason, continued administration of kava extract could enhance the metabolism of drugs that are substrates for this enzyme, possibly lowering the blood levels of prescription medications, thereby lessening their therapeutic effectiveness. In human volunteers, the daily oral administration of a relatively high dose of kava root extract resulted in significant inhibition of several endogenous drug metabolizing enzymes.[36] Because of the dose employed, it is unclear whether this finding has any clinical relevance. It is notable that this high dose did not cause liver toxicity, again suggesting that the toxic samples contained different constituents than those normally found in kava root extracts.

Consumption of high doses of kava extracts causes skin eruptions in susceptible individuals. This condition is characterized by patchy scales developing over large areas of the body. It generally resolves when kava extract use is terminated. The cause of this response is unknown, although it is thought to be an allergic reaction to some extract constituent.[3]

## Pharmacological Perspective

Historical use and contemporary clinical trials provide compelling evidence that extracts of kava root, if prepared properly and taken in appropriate quantities, are central nervous system depressants. This property is responsible for popularizing the use of this herbal supplement to relieve tension and anxiety, and to induce sleep. Evidence suggests that kavain, one of several kavalactones produced by the plant, may be most responsible for the pharmacological response to the extract, although other chemical constituents may contribute as well. The biochemical mechanism of the central nervous system depressant action of kava extract remains undefined, although studies point to a possible direct or indirect interaction with the GABA system in the brain. While there are reports of abuse of kava extract, as is the case with all central nervous system depressants, the liability for addiction, tolerance, and withdrawal appears low, even with continued use. It is noteworthy that when taken at recommended doses, the positive response to kava extract is not associated with the same degree of central nervous system side effects as is commonly observed with conventional depressants, such as ethyl alcohol and the benzodiazepines. Kava extract should not be consumed with other central nervous system depressants as the effects on alertness will be at least additive. Caution must also be exercised when taking kava extract while on any medication as kava constituents can affect the activity of some drug metabolizing enzymes, and thereby influence the response to other agents. In addition, there may be an additive, if not synergistic, toxic effect between kava and some drugs. This is particularly true if the kava extract is administered with other potentially hepatotoxic agents, such as acetaminophen, a common, over-the-counter analgesic and antipyretic. Most troubling are the reports of kava extract alone being associated with acute, irreversible liver toxicity. While the occurrence of this potentially lethal effect appears rare, and may be associated with the manner in which the extract is prepared, the possibility remains that some individuals may be prone to this response. Because of the rapid onset of this condition, and the lack of knowledge about its cause, it is impossible to predict who may be at greatest risk, or which extract preparations are most liable to provoke the response.

# 10

## Lavender (Lavandula angustifolia)

The fragrance of lavender was commonplace in ancient Greece. The Greeks rubbed the herb, which they called nardus or nard, on their skin for its restorative effects or added it to baths for its scent. The Romans are credited with popularizing the cultivation and use of lavender during their territorial conquests. In fact, its use can be traced back to biblical times, with lavender being described as a holy herb in the Old Testament. It is likely, therefore, that this plant has been appreciated since prehistoric times.

The qualities of lavender are so familiar that the word is used to denote a plant, a color, and an odor. While there are several species of lavender, Lavandula angustifolia, or English lavender, is the most favored because its aroma is particularly pleasing to humans. Other common species are spike lavender (L. latifolia) and French lavender (L. stoechas). The scent of lavender, the characteristic most responsible for its popularity, is variously described as floral, green, herbal, or woody.

As with many other scented plants, the ancients believed that lavender possessed curative powers. It was often a component of the elixirs of life that were concocted over the centuries. This includes mithridatum, a potion that dates to the second century BC.[1] While its constituents varied over the years, with up to 65 different herbals used to prepare mithridatum at any given time, this preparation was for two millennia a popular antidote for poisonings. In the sixteenth century, Friar Giovanni Andrea described "Elexir vite per conservare la humana natura," or the "elixir of life to preserve human nature."[2] French lavender was among the herbs in this mixture. It was thought that regular consumption of this brew would slow, and in fact reverse, the aging process. In the same century, Nicholas Culpeper, an English botanist and physician, wrote that the oil derived from English lavender was of particular value for treating all "griefs of the brain." He noted that it was so potent that only a few drops were needed for this purpose.[3]

Today lavender oil is used in cosmetics, as a flavoring for foods, in air fresheners, and for medicinal purposes. While in the modern era lavender oil has not been routinely employed as a systemic treatment for disease, preparations for its oral administration are now available. It is suggested these new products may be of value for treating anxiety, insomnia, and other conditions.

## Botany

Like lemon balm, lavender belongs to the mint family Lamiaceae. While the plant is believed to have originated in Asia, it is considered native in areas ranging from the Canary Islands, to Europe, Africa, the Middle East, and parts of India. Today lavender is cultivated and grows wild in temperate zones worldwide.

There are 39 species of lavender, along with numerous hybrids and cultivars. Only three species, all of which are found in Europe, have been used routinely as remedies: Lavandula angustifolia (English lavender), the closely related Lavandula latifolia (spike lavender), and Lavandula stoechas (French lavender). A fourth species employed for preparing lavender oil is Lavandula x intermedia. A hybrid of English and spike lavender, Lavendula x intermedia is native to areas where the parent species overlap, such as south central Europe. The hybrid is cultivated because its flowers produce greater quantities of the essential oil than English lavender. The essential oil of the hybrid is called lavandulin to distinguish it from that obtained from the parent species. The fragrance of English lavender is considered the most pleasing of the group. Lavender is a perennial shrub that from early June to late July produces flowers that are borne on spikes. The flowers range in color from white, to mauve, to purple.

## Therapeutic Uses

Because of their scent, lavender buds and the oil derived from them are used primarily as a food flavoring and for cosmetic purposes. There are lavender syrups, teas, and candies. Lavender oil is applied to the skin as a mosquito repellant. It is used as a fragrance in perfumes, creams, soaps, bubble baths, shampoos, and air fresheners.

In the sixteenth century John Gerard detailed the medicinal uses of English and spike lavender. He described them as remedies for heart palpitations, giddiness, and paralysis. According to Gerard, the benefits of this herb were realized by inhalation of the fragrance, bathing the temples in its oil, and by ingestion. Gerard mentions French lavender as being of value for treating headache, epilepsy, and the symptoms of stroke.[4] Lavender oil was also employed for its purported antiseptic and anti-inflammatory activities. It is placed in lotions, creams, and solutions for the topical treatment of insect bites, acne, muscle aches, sunburn, and lice. The evidence supporting the use of lavender for these conditions is mostly anecdotal.

There is growing interest in the use of lavender for its purported beneficial effects on the nervous system. This is not a novel idea. The eighteenth century *London Pharmacopeia*, a text describing the preparation and use of medical remedies, described a lavender tea for treating headache, fatigue, and exhaustion. In the early twentieth century, a Mrs. Grieve, who compiled extensive information on herbals, wrote in her *Modern Herbal* about a lavender-containing plant mixture that was taken to relieve exhaustion.[5] The results of contemporary laboratory animal and clinical studies suggest that lavender oil and some of its individual chemical constituents might be useful as treatments for pain, anxiety, insomnia, depression, and symptoms of dementia. For treating these conditions, lavender preparations are applied topically, by inhalation, or orally.[6]

## Constituents

The mature bud, or flower spike of the lavender plant is the primary source of the chemicals found in lavender products. While the plant leaves may occasionally be used for culinary purposes, the dried bud or bud extract is routinely employed as a food flavoring and for all other applications. Lavender oil is obtained by steam distillation of the flower spike to separate the essential oil from more water soluble components. The pleasant odor of the oil is due primarily to the presence of monoterpenes. Because the relative quantities of the individual chemical components vary during the maturation of the bud, the composition of lavender oil is dependent in part on the stage of development of the flower spike at the time of harvest.[7] The monoterpenes primarily responsible for the lavender fragrance are linalool and linalyl acetate. The concentration of these constituents varies widely among the various lavender species (see Table 10.1).[8] Moreover, linalool is produced by hundreds of different plants besides lavender. This includes flora as diverse as cinnamon, birch trees, rosewood, and citrus fruits.

**Table 10.1** Relative Concentration of the Major Chemical Components of Lavender Oil[a]

| | Species | | |
|---|---|---|---|
| **Chemical Component** | **L. angustifolia (English)** | **L. latifolia (Spike)** | **L.x intermedia (Hybrid)** |
| Linalool | 10-50[b] | 26-44 | 20-33 |
| Linalyl acetate | 12-54 | 0-2 | 19-26 |
| 1,8-cineole | 2-3 | 25-36 | 10 |
| Camphor | 0-0.2 | 5-14 | 12 |

[a] Adapted from Harborne and Williams[9]

[b] Values represent the percent of the component within this group of four chemicals.

While some of the variability in the chemical content of the different lavender species is due to growing conditions, the time of harvest, or the growth phase of the flower spike, genetic differences also play a prominent role. For example, it has been found that a mutant of Lavendula x intermedia produces only trace quantities of linalool, although this compound is abundant in the parent plant.[10]

There are 65 or more distinct chemicals in lavender oil, with linalool being the most abundant of these in the three plant species that are most commonly used to produce the oil (refer to Table 10.1).[9] Linalool represents up to 50% of this chemical group in English lavender, and one-third to nearly one-half in spike lavender and the hybrid. While the concentration of linalyl acetate approximates that of linalool in English lavender and the hybrid, it is much lower than linalool in spike lavender. Of the remaining two compounds, appreciable quantities of 1,8-cineole are present only in spike lavender, with camphor being only a small fraction of this chemical group in all three species (refer to Table 10.1). Because of their relatively high concentrations in the extract, not only the fragrance but the therapeutic benefits of lavender oil are thought to be mediated by linalool and linalyl acetate. For this reason, research has focused on characterizing the pharmacological properties of these two compounds. It remains possible that any of the 60 other chemicals in the distillate, either alone or in combination, may contribute to the pharmacological actions of lavender oil.

## Pharmacokinetics

Only a few pharmacokinetic studies have been performed on lavender oil and its constituents. This is presumably because lavender oil has historically been systemically administered either by inhalation of its vapors or by the oral consumption of the small quantities present in lavender teas or in food. Only recently has the oil been packaged and sold in a formulation suitable for oral administration of a fixed dosage for medicinal purposes. Assuming these products are popularized, it is anticipated that detailed pharmacokinetic studies will in the future be performed to define more precisely the extent of absorption, distribution, and metabolism of what are believed to be the pharmacologically active constituents in lavender.

Because lavender oil is contained in a number of liniments and cosmetic creams, there have been studies to determine whether linalool and linalyl acetate are absorbed through the skin. In vitro experiments suggest that both compounds make their way into deeper layers of the skin shortly after topical application.[11,12] The results indicate that linalool is more readily absorbed through skin than linalyl acetate.

In vivo studies of cutaneous absorption have been conducted with human subjects. Similar to the in vitro findings, these experiments indicate that linalool is absorbed through the intact skin. In one study, topical application of linalool was associated with a fall in blood pressure and skin temperature, suggesting a systemic response to this agent. These subjects did not display any signs of an increase in mood or in feelings of well being, indicating that the quantities absorbed were insufficient to affect the central nervous system.[13] Linalool is detectible in deeper skin layers within an hour of its placement on human skin. These levels declined within one to two hours after linalool was removed from the surface of the skin.[14] Evidence indicating that these compounds penetrate the skin completely is provided by the finding that both linalool and linalyl acetate are detected in blood following topical application of lavender oil.[15]

These data suggest that linalool and linalyl acetate, and perhaps other constituents of lavender oil, are sufficiently lipid soluble to cross the skin and accumulate in the systemic circulation. Assuming they are not rapidly metabolized in the intestine, these results suggest these compounds will readily cross from the gastrointestinal tract into blood when taken orally. It is not possible from these reports, however, to estimate the bioavailability, or fraction of an administered dose of these plant constituents, that will appear systemically, and whether either of these compounds penetrates into the brain in sufficient quantities to modify nervous system activity.

In vitro experiments indicate that linalool is metabolized by human liver enzymes.[16] One of the resultant compounds, 6,7-epoxylinalool, is thought to contribute to the skin irritation that some experience following the topical application of lavender oil. In vivo studies revealed that linalool was metabolically converted in rats to a number of different compounds following its continuous oral administration.[17] Consumption of linalool increased significantly the levels of certain drug metabolizing enzymes in rat liver. These findings confirmed the results of earlier studies with rats indicating that terpenoids, including linalool, when administered either by aerosol or orally, increase the quantity and activity of drug metabolizing enzymes.[18,19]

While few in number, these experiments suggest that linalool is extensively metabolized following systemic administration, and that long-term exposure to this lavender oil constituent may modify the rate and extent of metabolism of other agents, including prescription medications. Although these studies are too few in number to draw any definitive conclusions, they raise the possibility that any pharmacological response to lavender oil may be mediated by a metabolite of its chemical constituents, and that continuous consumption of this product could influence the activity of drugs being taken for other indications.

## Pharmacodynamics

Because inhalation of vapors containing lavender extract has been a popular mode of administration for decades, a number of laboratory animal studies have been aimed at defining responses following this type of exposure. Using mice it was found that inhalation of lavender oil vapor reduced locomotor activity, indicating sedation, and attenuated the excitation associated with administration of caffeine.[20] A correlation was found between these actions and the concentration of linalool in blood following inhalation of the lavender oil vapors. Others have reported that inhalation of certain concentrations of linalool alone, or of the oil, reduced anxiety in mice and rats, increased social interactions, and decreased conflict without causing a significant amount of sedation or memory impairment.[21-26] Sedation and memory dysfunction were both observed, however, following administration of higher doses of the oil or linalool. Of the compounds examined, linalool was found to be the most pharmacologically active of those contained in a lavender extract.[23]

Inhalation of either lavender oil or linalool is reported to suppress lipolysis by reducing the activity of the sympathetic nervous system.[27] This mode of administration also decreases plasma levels of norepinephrine, epinephrine, and dopamine.[28] These biochemical effects suggest a calming or sedating response to lavender oil and linalool.

Together, these studies indicate that pharmacologically active concentrations of lavender oil components may gain entry into the brain following exposure to vapors containing this herb. Because the nasal passages and lungs are highly vascularized, they allow for the rapid absorption of lipid soluble agents into the bloodstream. Alternatively, it is possible that the stimulation of olfaction by inhalation of lavender oil vapors indirectly affects behavior as the olfactory nerve is associated with the limbic system in the brain, an area involved in mediating emotions. A direct effect on the olfactory nerve would not require that significant amounts of linalool, or other lavender oil constituents, penetrate into the blood or brain following inhalation as their actions would be localized to the nasal passages.[29] Moreover, the pleasing fragrance and taste of lavender could, at least for some, be psychologically rewarding, thereby providing a relief from anxiety and an elevation in mood. Such a response would not necessarily require accumulation of lavender oil constituents in the brain.

It has been reported that injections of an aqueous extract of lavender attenuates learning deficits in an animal model of Alzheimer's disease.[30] As the learning and memory problems associated with this model are thought to be due to an inflammatory process, it is possible that the purported anti-inflammatory effects of linalool and linalyl acetate may be responsible for this response.[31,32] In contrast, others report that the injection or inhalation of linalool disrupts memory and learning in rats, suggesting this compound could worsen the symptoms of Alzheimer's disease.[25,33] These contradictory findings could indicate the presence of memory enhancing and memory inhibiting substances in lavender oil, with the memory blocking effect of linalool counteracted by some other component, or combination of chemicals in the extract. Alternatively, effects on memory may be dose-dependent, with certain amounts of linalool improving, and others inhibiting, memory consolidation in rats.

Other reported central nervous system effects of lavender oil or its constituents are anticonvulsant activity and analgesia. With regard to the former, systemic administration of linalool to mice protects against various types of seizures, displaying a profile similar to known antiepileptic agents.[34] Studies with mice and rats generally support the notion that lavender oil, as well as linalool and linalyl acetate themselves, are anti-inflammatory and analgesic in conventional animal tests for such activities.[35-37] These include assays that examine acute responses to a thermal or mechanical stimulus, as well as the more chronic discomfort associated with inflammatory pain. It has also been reported that linalool is effective in animal models of neuropathic pain.[38-40] Neuropathic pain results from damage to a nerve, is typically chronic, and is difficult to treat. As there have been many test compounds that were effective in animal tests for analgesia but that ultimately proved inactive in humans, the clinical significance of these findings with lavender oil remain to be proven. Because this herbal supplement has not historically been used to relieve pain it seems unlikely that it displays notable analgesic properties, at least when administered topically or by inhalation.

There are reports that linalool displays local anesthetic properties when applied directly to a nerve.[41,42] While such an effect is probably not clinically significant given its normal routes of administration, this finding could provide insights into its mechanism of action as an agent affecting central nervous system function.

Laboratory studies indicate that linalyl acetate displays antimicrobial activity. It is thought this effect is due to a linalyl acetate-induced increase in bacterial membrane permeability.[43] Another in vitro effect of linalyl acetate is relaxation of vascular smooth muscle.[44] While such an action could contribute to the drop in blood pressure reported after inhalation of lavender oil, data are lacking about whether this response occurs in vivo following administration of lavender oil, or of linalyl acetate alone. Oral administration of linalool to mice is reported to lower levels of blood cholesterol by inhibiting the synthesis of this lipoprotein.[45] This suggests that consumption of lavender oil-containing foods could have a beneficial effect on cardiovascular function.

Various clinical studies have been conducted to examine the responses of healthy and emotionally distressed subjects to inhalation of lavender oil or linalool vapors. Exposure to lavender oil vapors decreased stress levels in normal subjects and in individuals experiencing anxiety caused by the anticipation of an impending dental procedure.[46,47] Lavender oil aromatherapy was ineffective, however, in mitigating the anxiety experienced by those about to undergo a colonoscopy or esophagogastroduodenoscopy.[48] It is not known whether these conflicting findings reflect differences in the intensity of anxiety between these patient populations, in the chemical compositions of lavender oil preparation, or in the antianxiety response to this treatment.

Consistent with the findings from laboratory animal studies, inhalation of lavender oil or linalool alone causes sedation and improves sleep.[49,50] As these studies were conducted on healthy subjects, questions remain about the extent to which inhalation of these agents can benefit those suffering from sleep disorders. Nonetheless, the similarity between these results, those reported for laboratory animals, and the historic use of lavender oil as a sedative and hypnotic, suggests that this route of administration can, either directly or indirectly, modify central nervous system function.

Inhalation of lavender oil has been reported to improve somewhat the cognitive symptoms, and to decrease agitation, in some patients suffering from dementia and Alzheimer's disease.[51-53] In one study steps were taken, such as inclusion of appropriate placebo control groups, to ensure that any improvement was related to the herbal supplement rather than to other external variables, such as an increase in social interactions between the caregivers and the patient volunteers. The modest improvement noted and the fact that not all patients benefitted from the therapy suggest that the positive response is likely a reflection of the antianxiety and sedative effects of lavender oil, rather than to any specific action on the underlying neuropathology responsible for dementia and Alzheimer's disease.

Besides inhalation studies, clinical trials have been conducted using silexan, an oral formulation of lavender oil that is marketed in Europe.[54-56] One study entailed a double-blind comparison of silexan to lorazepam, a benzodiazepine, as a treatment for generalized anxiety disorder. Lorazepam is a drug commonly used to treat this condition. The results indicated that the daily consumption of the lavender oil-containing capsule was as effective as lorazepam administration in reducing the symptoms of this disorder. In a study of more than 200 patients it was found that oral administration of silexan decreased anxiety and improved the quality and duration of their sleep. When considering these findings, notice should be taken of the fact that two of the co-authors on these reports are employees of Dr. Willmar Schwabe GmbH & Co., the manufacturer of silexan. Although their participation does not prove a bias in analyzing and interpreting these results, definitive conclusions regarding the effectiveness of this product must await replication of these findings by others who have no financial stake in the product. While these clinical findings parallel those reported for the inhaled substance, no pharmacokinetic data are presented to demonstrate the presence in blood and the biological half-lives of lavender oil constituents following the oral administration of silexan. Such data are necessary to establish that the clinical effects reported are the result of a pharmacological response to the lavender preparation.

While several mechanisms have been proposed to explain the reported actions of lavender oil and its constituents on central nervous system function, most studies were aimed at examining their effects on GABA and glutamic acid neurotransmission. In vitro laboratory experiments suggest that the antianxiety effects of lavender oil may be due to the ability of linalool to enhance brain GABA receptor function in a manner similar to that reported for established anxiolytic agents.[57,58] However, others have been unable to detect an effect of linalool on GABA receptor activity.[59,60] Together, these results suggest that linalool probably does not interact potently and selectively with the receptors associated with this inhibitory neurotransmitter.

Various in vivo and in vitro studies have suggested the antianxiety, sedative, anticonvulsant, analgesic, and amnesic effects of lavender oil are due to blockade of glutamic acid receptors or to inhibition of the neuronal release of glutamic acid in the brain.[60-63] As the concentration of linalool necessary to block glutamic acid receptors is quite low, this mechanism seems plausible, especially because glutamic acid is known to play a role in mediating pain, anxiety, seizures, and memory.

Some studies suggest that chemical constituents of lavender oil can affect acetylcholine neurotransmission.[64] In particular, it has been reported that 1,8-cineole, a chemical found in the lavender oil extract (refer to Table 10.1), is a relatively potent inhibitor of acetylcholinesterase, the enzyme responsible for the destruction of acetylcholine. Blockade of acetylcholinesterase would be expected to enhance neurotransmission at acetylcholine synapses in brain and skeletal muscles. While such an effect might contribute to the reported beneficial effects of lavender oil in Alzheimer's disease, it would not explain the anticonvulsant and antianxiety effects of this herb. Moreover, if inhibition of acetylcholinesterase was the major action of lavender oil, numerous predictable side effects, including diarrhea and skeletal muscle twitches, should be associated with its use. As this is not the case, it seems unlikely that the concentration of 1,8-cineole in blood following administration of lavender oil is sufficient to inhibit acetylcholinesterase in a clinically meaningful way.

Studies of the direct action of linalool on nerve transmission suggest this compound is capable of blocking neuronal ion channels, an action typical for local anesthetics.[65,66] As the numbness that accompanies a local anesthetic effect is not generally associated with the use of lavender oil, even when it is applied topically, an action on ion channel activity probably does not contribute significantly to its reported pharmacological effects.

These mechanism studies indicated that, of the lavender oil constituents examined, linalool in particular may be capable of selectively modifying brain neurotransmitter systems in a manner that could account for the effects of this herbal supplement on central nervous system function. This conclusion is tentative, however, because information is lacking on whether linalool, or other chemicals contained in the lavender extract, attain and maintain the brain concentrations necessary to affect neurotransmission. Also, because of the lack of pharmacokinetic data it is unknown whether the doses and concentrations of lavender oil and its chemical constituents used in laboratory animal and in vitro studies bear any relationship to the amounts normally consumed by humans. This makes it impossible to attribute the central nervous system effects of lavender oil to any of the actions reported thus far on neurotransmitter systems.

## Adverse Effects

Side effects are not normally associated with the topical application of lavender oil, its inhalation, or oral administration. There was an anecdotal report suggesting that the topical application of lavender oil may have been responsible for gynecomastia, or breast enlargement, in three young boys.[67] However, a link between lavender oil treatment and endocrine changes in children remains unproven. There are also reports of skin allergies resulting from repeated application of lavender oil.[68,69] This allergic response, which is rare, is probably related to the presence in the preparation of oxidized forms of linalool and linalyl acetate.

With the availability of an oral preparation of lavender oil, there may be a greater possibility of an interaction with drug metabolizing enzymes, especially if the herb is taken over an extended period of time. The laboratory animal findings indicate that lavender oil constituents are extensively metabolized and can alter the quantity and activity of drug metabolizing enzymes in the liver.[17-19] For these reasons, caution should be exercised when taking this herb in combination with conventional medications as the lavender oil constituents may increase or decrease the effectiveness of other agents. Alternatively, continuous administration of prescription or over-the-counter drugs that interact with the same liver enzymes as lavender oil components could affect the response to this herb, making a given dose more or less active than anticipated.

## Pharmacological Perspective

There is little doubt that lavender oil affects central nervous system function, either directly or indirectly. At a minimum, its odor and taste are perceived as pleasant by most and as calming and sedating by many. These conclusions are based on the historical record of its use, as well as laboratory and clinical studies of its effects on human and rodent behaviors. What is uncertain is whether these reported responses to lavender oil are due to a selective pharmacological effect on the brain by one or more of its chemical constituents, or represent psychological responses to its pleasing fragrance and taste. As some of the reports on its central nervous system actions, especially in laboratory animals, involved administration in ways that preclude a response to odor or taste, there is circumstantial evidence supporting the hypothesis that lavender oil contains pharmacologically active ingredients. If so, these data indicate that linalool and linalyl acetate, two of the more abundant extract constituents, are likely candidates for mediating some, if not all, of the central nervous system responses to this herb. What is sorely lacking are pharmacokinetic data demonstrating directly that lavender oil constituents, or their pharmacologically active metabolites, are present in blood at concentrations, and for a sufficient period of time, to account for the clinical effects of this herbal supplement. This information is essential for demonstrating that lavender compounds appear in the bloodstream and accumulate in the brain. Until such data are forthcoming, it is appropriate to reserve judgment on its clinical utility for affecting central nervous system function.

# 11

## Kudzu (Pueraria lobata)

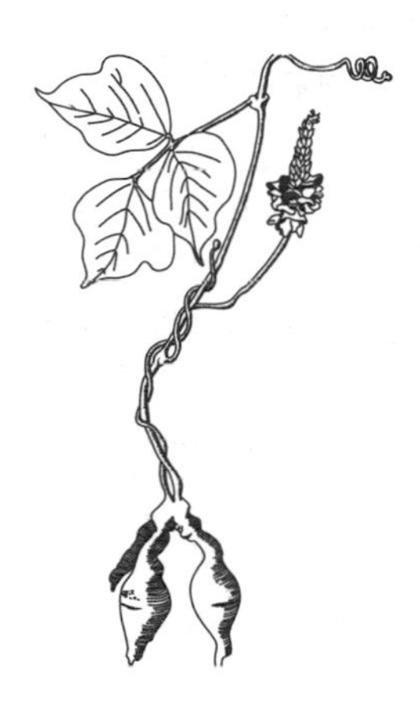

A perennial, kudzu is a rapidly growing, climbing vine. Native to southern Japan and southeastern China, the name is derived from Kuzo, the Japanese designation for this species. In China the root preparation is referred to as Ge Gen. The plant was first introduced into the United States by the Japanese during the 1876 Philadelphia World's Fair. It has been referred to as porch vine because of its ability to climb a trellis quickly and produce shade. Other informal appellations include "foot-a-night vine" and "the vine that ate the South," which also refer to its rapid and aggressive growth. Its cultivation in the United States was encouraged in the first half of the twentieth century for use as fodder and to reduce soil erosion. As the climate in the southeastern United States is ideal for its growth and propagation, kudzu spread extensively throughout the region and is now considered by many to be a noxious weed. The density of kudzu vine can be great enough in forests to prevent tree growth.

Kudzu has for centuries been employed for a number of uses. These range from soups, lotions, teas, and juices to fabrics, starches, and furniture. It is now being recommended as a biofuel. Kudzu flowers, leaves, and roots have for some time been consumed in Asia for treating a variety of ailments, including cancer, headaches, cardiovascular conditions, respiratory problems, hay fever, psoriasis, migraine headaches, and diabetes. With regard to central nervous system disorders, more than 1,000 years ago the Chinese reported its possible value as a treatment for alcoholism and for the symptoms of alcohol hangover. Pharmacological studies have yet to demonstrate whether this purported benefit is due to a direct effect on the brain or is secondary to an effect on alcohol absorption or metabolism. More recent work suggests kudzu may have neuroprotective properties. If so, it may be a potential therapy for minimizing the brain damage associated with a stroke, and perhaps in slowing the brain cell death responsible for Alzheimer's, Parkinson's, and other neurodegenerative disorders. Given these possibilities, a critical assessment of the pharmacological properties and central nervous system effects of kudzu is warranted.

## Botany

As the aerial parts of the kudzu plant die in cold weather and prolonged freezing can be fatal for the roots, this plant is most commonly found in warmer regions, such as the southern United States. While kudzu is native to much of the southern Asian continent and associated islands, fossil plants of the genus Pueraria are also located in temperate regions of Eurasia. Diversification of the Pueraria species throughout Asia and Oceania appears to have begun in the mid-Miocene epoch, or about 12 million years ago.[1] The Miocene was characterized by a relatively warm global climate.

Kudzu reproduces by either seed or vegetative means. With the latter, the vine shoots spread horizontally from the root crown. During the growing season portions of the root system swell and form edible, starch-filled tubers. These tubers are sometimes ingested for medicinal purposes.

There may be up to 17 species of kudzu, although its taxonomy is a matter of debate. In one system there is only a single name for kudzu, Pueraria lobata. In another, the species lobata is one of three varieties of Pueraria montana. The other two are Pueraria chinensis and montana. In this system, Pueraria montana (Lour.) Merr. var. lobata (Willd.) Maesen and Almeida is the scientific designation for the cultivar found in the United States.[2] The taxonomic classification does not, however, reflect any differences in plant constituents believed to be responsible for therapeutic activity. All species produce the same substances. Quantitatively, however, their concentrations vary within, as well as between, cultivars.

The genus Pueraria belongs to the legume family, Leguminosae. The Leguminosae encompass a variety of plants, including common peas and beans. Their characteristic flower is well-designed for insect pollination. Both the soy bean (Glycine max) and kudzu belong to Glycininae, a subgroup of the Leguminosae. All members of this family produce isoflavones, chemicals thought to be responsible in part for their therapeutic benefits.

## Therapeutic Uses

Kudzu has for centuries been a staple among Chinese herbal medicines. The sixteenth century manual of medicinal herbs by Shih-Chen Li recommends the root be used for its antipyretic and antiemetic actions, and as a treatment for poisoning. Kudzu seeds and flowers were prescribed for ailments associated with alcoholic excess, and the leaves applied to wounds to facilitate healing.[3]

A treatise on modern Chinese herbal therapy includes a list of recommended uses for kudzu root. Because it reportedly decreases vascular resistance, it is recommended as a treatment for hypertension and for increasing blood flow to the heart and brain.[4] The vascular effect might also explain its reported efficacy in minimizing tissue damage associated with cardiac and cerebral infarcts. Kudzu is also recommended as a treatment for cancer, for osteoporosis, and migraine headaches. Recently there has been a renewed interest in the historic use of kudzu as a remedy for hangover and alcoholism. This has led to a number of published reports on the effectiveness of kudzu preparations, and individual isoflavones isolated from the plant, in reducing alcohol consumption and the symptoms associated with alcohol withdrawal.[5,6] Included among these are studies on possible biochemical effects of kudzu that may be responsible for its various therapeutic actions.

## Constituents

As with all plants, kudzu flowers, leaves, and roots contain scores of different chemical agents. To date, those in kudzu thought to be the most bioactive are certain isoflavones and their glycosides.[7] A glycoside is a noncarbohydrate molecule bound to a sugar moiety. In this case, the isoflavone is the noncarbohydrate portion of the molecule. The term aglycone is used to describe the noncarbohydrate portion of the molecule after removal of the sugar moiety. More than 20 different isoflavone glycosides and aglycones have been identified in kudzu extracts. Those generating the greatest interest as possible therapeutics are the glycosides puerarin, daidzin, and genistin, and the aglycones daidzein (see Figure 11.1) and genistein.

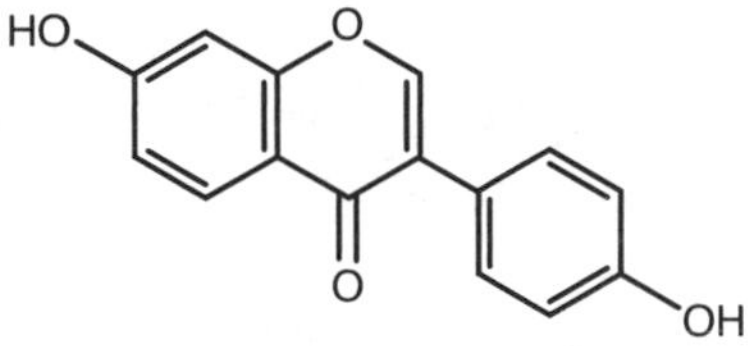

Figure 11.1 Chemical structure of daidzein (Wikipedia)

The relative percentages of the various isoflavones and isoflavone glycosides vary in different cultures of kudzu.[8] In the root of the Pueraria lobata found in the United States, puerarin is present in higher concentrations than the other isoflavone glycosides and the aglycones.[9] A study of 96 samples of kudzu collected at different times of the year revealed that the roots of three-year-old plants harvested in January have the highest content of isoflavonoids.[10] The degree of variation adds to the difficulty in ensuring consistency of content in kudzu products sold commercially. It also complicates interpretation of laboratory results obtained using unpurified plant extracts or powders. As the kudzu isoflavones have been chemically synthesized, it is possible to obtain pure samples for studying the pharmacological properties of these compounds. This makes it much easier for laboratories to reproduce findings by eliminating the variability that accompanies the use of extracts or plant products from different cultivars that may have been harvested at different times of the year.

As a group, the kudzu flavonoids are considered phytoestrogens. This term is used to describe plant products that display estrogenic activity. Estrogen is the primary female sex hormone. While structurally different from estrogen, phytoestrogens are capable of stimulating or blocking estrogen receptors. Given this property, such agents can induce or inhibit estrogenic effects. Such actions could explain, in part, some of the medicinal benefits attributed to kudzu.

## Pharmacokinetics

Studies have been conducted on the absorption, organ distribution, and metabolism of the chemical constituents in kudzu extracts, with particular emphasis on the isoflavone components.[11-18] The results vary depending on the animal species studied, route of administration, source of the extract, and method of analysis. In general, the data indicate that the kudzu isoflavones, both the glycosides and aglycones, are absorbed to a limited extent following oral administration and are rapidly eliminated. For example, blood levels of puerarin in the rat following oral administration are maximal in less than an hour, with the half-life for elimination being less than two hours.[18] In humans, however, the time for attaining maximum blood levels ranged from five to nine hours for the various glycosides and aglycones, with the half-lives for elimination averaging six hours.[12] Absorption appears to be due to both passive diffusion and carrier-mediated transport.[15] It has also been found that some of these isoflavones are metabolized by intestinal bacteria to agents displaying estrogenic activity, such as equol, and that the parent isoflavone may be converted to other metabolic products in their initial passage through the liver.[12,17] This rapid biotransformation is one reason for their relatively low and variable bioavailability.

Attempts have been made to enhance absorption of certain isoflavones by modifying the vehicle used for administration of purified substances.[19,20] These approaches, which include use of phospholipid complexes or nanoparticles, tend to increase the extent of absorption and the isoflavone concentrations in various organs.

The most precise pharmacokinetic data on the chemical components of kudzu are obtained using chemically pure isoflavones. Plant extracts, such as those contained in consumer products, are less useful for such studies given the high degree of variation in the concentration of constituents among these preparations. This variability makes it difficult to know the amount of any particular chemical being administered. On the other hand, the use of pure chemicals makes it impossible to know whether other components of the extract influence the rate and extent of absorption, or the metabolism, of individual constituents. Because of this, results with pure compounds may not in all cases accurately reflect their pharmacokinetics when they are taken along with the other plant constituents, as is the case with commercial products. In one study a methanol extract was administered to mice, with body fluid measurements indicating the isoflavones were rapidly absorbed and subsequently eliminated to a large extent in urine and feces. Organ tissue analysis 24 hours after administration revealed these compounds were widely distributed throughout the body with the highest concentration in the liver.[21] As none of these compounds was detected in the brain, it was concluded that puerarin, daidzin, and malonyl daidzin do not penetrate into the central nervous system, or are rapidly eliminated from this region of the body. Such data call into question whether isoflavone components of kudzu extracts directly affect the brain.

Qualitatively similar results were found when studying the absorption, distribution, metabolism, and elimination of puerarin when the purified compound was administered alone to rats.[22,23] These experiments revealed that puerarin was absorbed into blood following oral administration, reaching its maximum serum concentration within approximately 40 minutes, declining thereafter. In this case the highest concentration of compound was found in the lungs, although some was detected in the brain as well.

The absorption in rats of a mixture of the kudzu aglycones daidzein, genistein, and glycitein has been compared with the absorption of daidzin, genistin, and glycitin, their corresponding glucosides.[24] The results indicate that the oral bioavailability of these agents varies between 8% and 35%, with no consistent differences noted between the glycosides and the aglycones. These data indicate that these isoflavones appear in blood following oral administration, although the bioavailability is limited. The differences between the findings with the pure substance as compared to an extract highlights the difficulties associated with comparing the pharmacokinetic properties of an agent administered alone to when it is given as part of a compound mixture. Inasmuch as kudzu extract is most commonly taken by consumers, results with mixtures more accurately reflect what occurs with normal usage. Such data are inconclusive as to which, if any, of the kudzu isoflavones or their metabolites penetrate into the brain in sufficient quantities and for a sufficient period of time to have any clinically significant effect on central nervous system activity.

## Pharmacodynamics

Over the centuries, kudzu has been recommended for treating a host of conditions from cancer to migraine headaches and diabetes. In the ancient literature there is limited mention of this plant preparation having any obvious effect on central nervous system function, and no suggestions it may be of value in treating neurological or psychiatric disorders. Kudzu has, however, been reported to be effective as a treatment for alcoholism and the symptoms associated with alcohol withdrawal. The use of kudzu for these purposes has been documented for more than 1,000 years. A kudzu-induced reduction in alcohol consumption could be due to an effect on the brain that modifies the action of alcohol on the nervous system, or could be the result of an effect of kudzu constituents on the absorption, distribution, or metabolism of alcohol, thereby reducing the amount of ethanol that reaches the brain. In the latter case the kudzu-derived compound would not have to penetrate into the central nervous system to be effective.

It is possible that the ancient Chinese kudzu preparation had somewhat different constituents than those used today. This would make it difficult to draw firm conclusions about the possible beneficial effects of contemporary preparations on the basis of reports originating centuries ago in a different culture. For example, while the early Chinese used preparations of the kudzu flower to treat alcoholism, most products sold today are extracts prepared from the roots. Inasmuch as a metabolite of the isoflavone tectoridin, which is localized to the kudzu flower, protects mice from ethanol-induced liver toxicity, it is possible that the beneficial effect in treating alcoholism is peculiar to the flower preparation.[25,26] Nonetheless, there has been a renewed interest in the possible use of kudzu to treat alcoholism and alcohol withdrawal, with studies performed to determine whether root constituents have such an effect and, if so, to define the mechanism of action for this response. Laboratory animal studies show that administration of kudzu extracts, or of purified puerarin or daidzin, isoflavones glycones present in these root extracts, lessens the anxiety associated with alcohol withdrawal and reduces alcohol intake in rats and hamsters conditioned to consume this beverage.[6,27-29] In one study an attempt was made to measure brain levels of puerarin in alcohol-preferring rats after administration of the purified isoflavone at a dose that reduces voluntary alcohol intake by 50%.[30] No puerarin was detected in the brain even though its administration had a significant effect on alcohol consumption. This suggests that the antialcohol effect of puerarin, and of possibly the kudzu extract as a whole, is due to an action outside of the central nervous system.

Clinical trials of kudzu root extract have yielded conflicting results with regard to its utility as a therapy for alcoholism. In one pilot study the consumption of 1.2 g of kudzu extract twice daily for a month had no effect on the sobriety or alcohol craving of chronic alcoholics.[31] In contrast, a later study found that administration of kudzu root extract for seven days significantly reduced voluntary alcohol consumption in heavy drinkers.[32] The differences in these findings could be due to the fact that different populations of patients were examined, with in one case the subjects being classified as chronic alcoholics, whereas in the other they were characterized as heavy drinkers. Another explanation could relate to differences in the chemical composition of the kudzu preparations used in these studies, as variations would be anticipated among cultivars and manufacturers. In either case, the effectiveness of kudzu as a treatment for alcoholism remains a matter of debate.

Studies aimed at defining the possible mechanism of action of kudzu in reducing alcohol consumption and withdrawal symptoms focus on the fact that the isoflavones derived from this plant, in particular daidzin, inhibit aldehyde dehydrogenase-2, an enzyme involved in the metabolism of ethanol and various other chemical substances (see Figure 11.2).

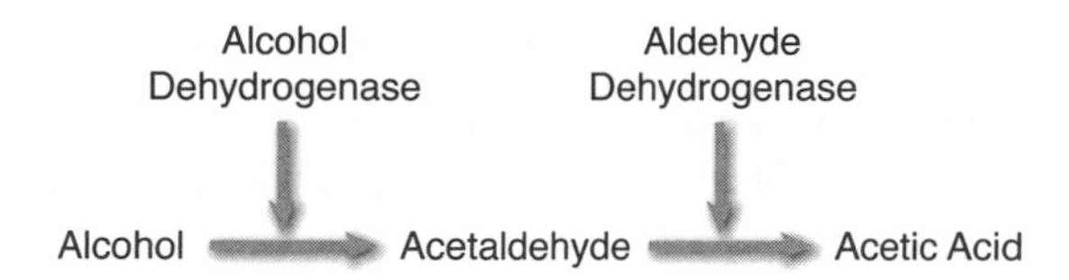


Figure 11.2 Metabolism of alcohol

As shown, in humans alcohol is converted to acetaldehyde by the enzyme alcohol dehydrogenase, with the acetaldehyde then metabolized to acetic acid by aldehyde dehydrogenase. Many of the symptoms associated with a hangover, including headache and nausea, are due to the accumulation of acetaldehyde in the blood, which can be significant if a large amount of alcohol is consumed over a short period of time, or the enzyme aldehyde dehydrogenase is inhibited. Disulfuram, a drug used for the treatment for alcoholism, is an inhibitor of this enzyme. Because a patient taking disulfuram knows he will become quite ill if even a small amount of alcohol is consumed, he is motivated to resist the temptation to consume alcoholic beverages. In this way, disulfuram can be helpful in modifying the behavior of an alcoholic.

In 2000, daidzin was found to be a selective inhibitor of aldehyde dehydrogenase-2.[33] As compared to other dehydrogenases in this family, aldehyde dehydrogenase-2 contributes in only a minor way to the metabolism of ethyl alcohol, although it is important in the metabolism of certain endogenous chemicals, such as serotonin, norepinephrine, and dopamine.[33-35] This explains why administration of kudzu extracts or isoflavones at doses that reduce alcohol consumption, does not, like disufuram, a more nonselective inhibitor of this enzyme family, increase blood levels of acetaldehyde. As studies with chemical derivatives of daidzin indicate a correlation between inhibition of aldehyde dehydrogenase-2 and reduction in alcohol consumption, the two actions may be related.[36,37] While there are suggestions that the kudzu isoflavones act in the brain to modify the response to alcohol,[38] the current weight of evidence indicates that these isoflavones probably act outside the central nervous system to modify the production of endogenous agents, which in turn affect the absorption, distribution, or action of alcohol in such a way that its consumption is less rewarding. It is unknown whether this same action is responsible for the reported beneficial effect of kudzu in reducing the symptoms of alcohol withdrawal.

While it does not appear that the effect of kudzu constituents on alcohol consumption is mediated through a direct action on the brain, there are reports that phytoestrogens, such as daidzin and chemically related agents, influence cognition, have neuroprotective properties, and reduce anxiety.[6,39-42] The extent to which these actions are due to a direct effect of the isoflavones or their metabolites on the brain is unknown.[42]

Outside of possible central nervous system actions, puerarin, a potassium channel inhibitor, has been proposed as a treatment for cardiac arrhythmias. In addition, the kudzu isoflavones genistein and daidzein restore vascular function in spontaneously hypertensive rats, confirming the potentially beneficial cardiovascular actions of these compounds.[43] A number of studies also demonstrate the anti-inflammatory effects of these isoflavones, in particular daidzein and genistein.[44,45] This has led to the suggestion that these agents, and perhaps other isoflavones or their derivatives, may be useful for the treatment of asthma and rheumatoid arthritis. There are also suggestions that daidzin and genistein inhibit breast cancer cell motility and decrease the risk of lung cancer.[46,47] Although some of the anti-inflammatory effects of these compounds, such as inhibition of NF-kappaB and tissue necrosis factor-α, could explain such effects, it remains to be seen whether these in vitro actions occur in humans at clinically appropriate doses.

## Adverse Effects

Clinical and laboratory animal studies suggest that neither the use of kudzu root preparations nor any of the individual isoflavones studied thus far is associated with significant adverse effects. This is not surprising inasmuch as the isoflavones found in kudzu are common to the legume family and therefore widely consumed as food products. For example, soy beans are a major source of genistin in the diet.

Studies have suggested that consumption of kudzu isoflavones for as little as ten days modifies the activity of drug metabolizing enzymes and intestinal drug transporters.[48-50] Other concerns, based on in vitro studies, include a possibly adverse effect on diabetics secondary to a daizdin-induced inhibition of glucose transport, and the accumulation of acetaldehyde or other substances in tissue as a result of the long-term inhibition of aldehyde dehydrogenase-2.[51,52] Overall, the data accumulated to date suggest that kudzu and its isoflavones are safe for human consumption at the recommended doses.

## Pharmacological Perspective

Kudzu root extract, and individual isoflavones contained therein, suppress excessive alcohol consumption in laboratory animals and may ameliorate alcohol withdrawal symptoms. While these effects have been noted for centuries, the human data are mostly anecdotal, with clinical trials yielding conflicting findings. Inasmuch as administration of kudzu has no obvious central nervous system consequences, it seems unlikely that an effect on alcohol consumption and withdrawal is due to an action on the brain. Rather, it appears the isoflavone constituents, and perhaps other chemicals in the plant product, modify enzymes responsible for the production of endogenous chemicals that can in turn influence the pharmacokinetics of alcohol or its central nervous system action. There is current interest in a possible role of inflammation in a variety of disorders, including depression, cardiac insufficiency, and cancer, in addition to the recognized role in arthritis. The anti-inflammatory and antioxidant actions of isoflavones include inhibition of NFkappaB, a cellular agent that initiates the inflammatory cascade. Both daizdin[53] and puerarin[54] have been shown to inhibit NFkappaB. Because kudzu and its constituent isoflavones appear safe, they would be superior to other drug-based approaches for treating alcoholism if further clinical studies show they are effective in this way.

Based on the data accumulated thus far, the value of kudzu as a neuroprotectant or anxiolytic is even more speculative than its possible utility as a treatment for alcoholism. This is because there is evidence indicating that the isoflavones may not penetrate into the brain, which is generally necessary for an agent to display neuroprotectant or anxiolytic activity. Also, because many of the kudzu constituents are rapidly and extensively metabolized in the intestine and liver, it may be difficult to identify the chemicals in the extract, or their metabolites, that may directly influence the brain. Until such substances have been characterized, and their accumulation and retention in the brain demonstrated, it would appear that any responses attributed to kudzu are due primarily to effects outside the central nervous system.

# 12

## Daffodil (Narcissus pseudonarcissus)

It has been known for centuries that portions of the daffodil plant, the bulb in particular, affect central nervous system function. However, because it causes nausea and vomiting, daffodil is generally not taken orally. With the relatively recent discovery that a daffodil constituent is useful in the management of Alzheimer's disease, this plant is an example of one that yields an agent useful for modifying central nervous system function even though the herb historically has not been used for that purpose. The clinical utility of daffodil also illustrates the value of alkaloids, a class of compounds produced by many plants. In fact, more alkaloids have been developed as prescription medications, or employed recreationally for their central nervous system effects, than flavonoids, the other major class of plant products. The discovery that daffodil produces a potent compound that can beneficially affect central nervous system function also demonstrates that plants are still a valuable source of new therapeutics.

The clinically active component of daffodil is the alkaloid galantamine, also known as galanthamine. Alkaloids are produced by plants, fungi, bacteria, and animals. The biochemical pathways for synthesizing these agents evolved to protect the plant from predators, as these compounds are often toxic. Thanks to its alkaloids, daffodil thrives as a species because its taste is unappealing and its consumption is potentially lethal to insects and others that may be interested in making it part of their diet. The alkaloid content of daffodil, and of other plant species, is generally quite small. However, because of their pharmacokinetic and pharmacodynamic properties, many alkaloids can, following oral consumption, readily accumulate in the brain where they interact with neurotransmitter systems. Because of their potencies, only small quantities are needed to have a significant behavioral effect. This explains why daffodils are so commonly and consistently seen sprouting in the same area of the lawn each year, whereas the appearance of other plants varies as a function of the local population of rabbits and other predators.

Plant alkaloids are a rich source of prescription and recreational drugs. Numbered among this chemical class are caffeine, nicotine, morphine, cocaine, quinine, and atropine. Mescaline, which is derived from a cactus, and psilocybin, which is produced by certain mushrooms, are hallucinogenic alkaloids that humans have used for centuries for religious and recreational purposes. This partial list of behaviorally active alkaloids illustrates the therapeutic potential of such compounds. Further evidence is provided by written records indicating that plants containing biologically active alkaloids have for more than 4,000 years been used as herbal remedies and poisons. The discovery that a common plant, such as daffodil, produces a clinically useful alkaloid is therefore not surprising. The fact that it took several millennia to appreciate its medicinal value is testimony to the difficulties sometimes associated with identifying a biologically active plant product with the chemical properties needed to be a useful drug.

The origin of the narcissus designation for this genus is a matter of dispute. Some claim the name derives from Narcissus, a character in Greek mythology. According to legend, because of his vanity and pride Narcissus showed only disdain for others. To punish him for such behavior, Nemesis, the Greek goddess of retribution, conspired to have Narcissus see his own reflection in a pond. He immediately fell hopelessly in love with what he saw. He was so transfixed with himself that he remained staring at the image until, according to one version, he died from starvation. It is said that narcissus plants first sprouted on the spot where he died. Others contend the plant name comes from the Greek word *narcoun*, which means to benumb. Narcoun is also the root for the term *narcotic*. According to this concept, the plant was called narcissus because consumption of the bulb depresses central nervous system function, not because of any relationship to Narcissus. There is more agreement on the origin of the word daffodil, the common name for Narcissus pseudonarcissus. It is thought that daffodil is a corrupted version of the Greek word *asphodel*, a flowery plant in the lily family. Regardless of its etymology, narcissus has been, and remains, a popular plant because of the beauty of its flower and ease of cultivation. Therapeutic value can now be added to this list of positive attributes.

## Botany

In his *Enquiry into Plants*, Theophrastus, a fourth century BC Greek botanist and philosopher, mentioned three different types of wild narcissus. These are now known as Narcissus serotinus, Narcissus poeticus, and Narcissus tazetta.[1] Two thousand years after Theophrastus, the English herbalist John Gerard categorized narcissus plants into two groups: narcissus and pseudonarcissus.[2] The latter is now referred to as daffodil. The genus Narcissus is a member of Amaryllidaceae, a family of plants noted for producing a unique series of alkaloids. More than 80 chemically related alkaloids are manufactured by plants of this genus.[3] Narcissus plants, of which there are some 50 to 100 species, originated in parts of southern Europe, Asia, and North Africa. While most blossom in the early spring, a few species bloom in autumn. A hardy plant, the daffodil is widely distributed, being found most commonly in damp woodlands and grassland regions throughout the world.

Amaryllidaceae is one of the families of Monocotyledonae. Other members of this clan are Liliaceae (lilies) and Graminae (grasses). As with many members of Amaryllidaceae, the narcissus plant develops from a scented bulb. The flowers may be solitary or grow as a cluster on the primary flower stalk. The bell-shaped flower has six floral leaves surrounding a corona. While most daffodil flowers are yellow to gold in appearance, there are hybrids that display a range of colors.

Most species of daffodil flower in March or early April. The plant grows to approximately one foot in height.

The discovery that daffodil is a source of galantamine increased interest in developing horticultural methods for producing bulbs with greater concentrations of this alkaloid to increase their commercial value. Significant progress has been made in this regard.[4]

## Therapeutic Uses

Daffodil has long been appreciated for the beauty of its flowers and its medicinal properties. Pliny, a first century Roman naturalist, listed two major uses for narcissus. One was as an emetic and the other as a topical treatment for wounds and sprains.[5] Echoing Pliny, Gerard described a plaster made by mashing the bulb with honey or animal fat. The resultant mixture was applied to wounds, burns, and sprains, and used for the topical treatment of joint pain.[2] He noted that when taken orally the bulbs caused vomiting. He also described a drink made from ground and strained bulbs as a treatment for cough and colic.

There are records of daffodil bulb and flowers being taken for their central nervous system effects. The conditions treated included hysteria and epilepsy. It was once believed that consumption of dried daffodil flowers could relieve lung inflammation and dysentery. A narcissus bulb oil extract was recommended as a cure for baldness and as an aphrodisiac. Today, a narcissus bulb tincture is used as a homoeopathic remedy, and a bulb extract liniment is available for treating muscle aches and pains. The daffodil alkaloid galantamine is taken by some to help achieve lucid dream or out-of-body experience, to facilitate sleep, or to improve memory consolidation while sleeping.

The major therapeutic value of narcissus bulbs is as a source for galantamine, which was first isolated from the Caucasian snowdrop, Galanthus woronowi. Employed since the 1950s as a medication in Eastern Europe, galantamine was first used in that part of the world as a treatment for polio and other nervous system disorders, including myasthenia gravis and asthenia. Today it is prescribed to treat some of the symptoms of Alzheimer's disease. Because galantamine is one of the major alkaloids found in daffodils, their bulbs are a source for the commercial production of this product.

## Constituents

All parts of the daffodil plant are poisonous and potentially toxic to humans and other animals. Even topical application of plant constituents is known to cause central nervous system and cardiovascular dysfunction in susceptible individuals. Of the many chemicals present in daffodil, the alkaloids are of greatest pharmacological and toxicological interest. This is not to imply that there are no other chemicals of pharmacological value in daffodil extracts, only that such compounds have yet to be identified and characterized. The four major daffodil alkaloids are galantamine (see Figure 12.1), lycorine, galanthine, and haemanthamine. Of the 80 or so alkaloids produced by plants in the narcissus genus, more than 20 are found in daffodil cultivars.[3]

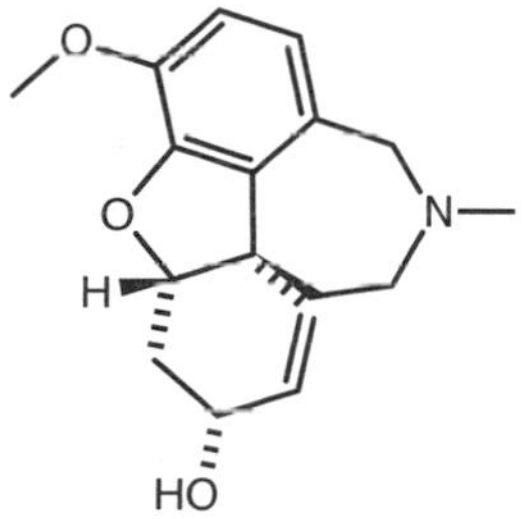

Figure 12.1 Chemical structure of galantamine (Wikipedia)

Galantamine was first extracted, its chemical structure identified, and its biochemical effects characterized in 1951.[6] The product sold today is either produced by chemical synthesis or harvested from plant material. It was discovered almost immediately that galantamine is an inhibitor of acetylcholinesterase, the enzyme responsible for the destruction of acetylcholine, an important neurotransmitter in the brain and other organs.

Lycorine was first identified in the spider lily, Lycoris, a member of the Amaryllidaceae family. The most common of the alkaloids produced by Amaryllidaceae, lycorine is an emetic, an anti-inflammatory agent, and an expectorant. Its ability to increase salivation is probably the reason it was used by the ancients as a treatment for cough. As an emetic, daffodil bulb was a popular therapeutic when it was believed that vomiting helped rid the body of the excess "humors" that were thought responsible for illnesses. While research suggests that two of the daffodil alkaloids, lycorine and haemanthamine, halt the proliferation of certain types of cancer cells, further work is needed to define the clinical significance of this finding.

## Pharmacokinetics

Galantamine is rapidly absorbed from the gastrointestinal tract. In rats, 77% of an orally administered dose appears in the bloodstream. Its elimination half-life, a measure of how long the compound remains in the body, is four to five hours. Following the oral ingestion of a single dose, the concentrations of galantamine were found to be greatest in rat liver and kidney, followed by the salivary and adrenal glands.[7] Significant concentrations of galantamine are present in the brain after intravenous administration to mice, indicating this substance is capable of accumulating in the central nervous system.[8] These pharmacokinetic studies suggest that galantamine readily passes from the gastrointestinal tract into blood and from the blood into other tissues, including the brain. The ratio of the galantamine concentration in brain as compared to blood indicates it accumulates and is retained in the central nervous system in quantities sufficient to induce pharmacological effects.

The results of human pharmacokinetic studies are similar to those reported for rats. In human subjects, galantamine quickly appears in the general circulation following oral administration. The elimination half-life is approximately five hours, with the urinary excretion of galantamine metabolites being complete within 72 hours of consumption.[9,10] It is estimated that 25% of the dose is excreted unchanged, with the remainder eliminated as metabolites.[11] This is consistent with the finding that a large portion of the administered agent is absorbed into the systemic circulation from which it can pass into the brain.

As for lycorine, approximately 40% of an orally administered dose is absorbed into the bloodstream of dogs.[12] Its half-life in blood is only 15 minutes. In vitro studies suggest that lycorine is a potent inhibitor of some drug metabolizing enzymes.[13] This action is probably of little clinical importance as neither daffodil extracts nor lycorine alone are normally administered systemically, especially for prolonged periods of time. Unless significant amounts of lycorine are absorbed through the skin following the continuous topical application of a daffodil extract, this compound is unlikely to interfere with the action of other medications.

## Pharmacodynamics

In vivo and in vitro laboratory studies conclusively demonstrate that galantamine augments the acetylcholine neurotransmitter system. It accomplishes this in two ways. One is to prolong the duration of action of acetylcholine by inhibiting acetylcholinesterase, the enzyme that destroys this neurotransmitter.[6] Galantamine also enhances the responsiveness of the acetylcholine nicotinic receptor system.[14-16] As acetylcholinesterase and nicotine receptors are located in the brain, as well as other organs, these actions explain the beneficial effect of galantamine on central nervous system function, as well as the side effects associated with its use.

The acetylcholine system plays a key role in memory retention and recall. This is demonstrated by the fact that drugs that block acetylcholine receptors, such as the alkaloids atropine and scopolamine, cause amnesia for events that occur while an individual is under their influence. The finding that galantamine is capable of reversing in laboratory animals the memory deficit associated with scopolamine administration confirms that the effects of this daffodil constituent are related to its ability to facilitate acetylcholine neurotransmission.[17]

Dementia is characterized by progressive memory loss, learning deficits, problems with judgment, and emotional lability. While dementia can be due to many causes, in the aged population it is commonly associated with Alzheimer's disease. Because the memory impairment, especially for recent events, is similar to that occurring following administration of an acetylcholine receptor antagonist, it was thought that it may be due to a deficiency in acetylcholine transmission. Studies with human autopsy material confirmed this suspicion by demonstrating that Alzheimer's disease is characterized by a significant loss of cholinergic neurons in certain brain regions. This includes the hippocampus, a limbic brain area known to be important for memory consolidation and recall. For this reason, efforts were made to treat Alzheimer's disease with drugs that improve acetylcholine transmission.[18] Several such agents are approved for this use. This group includes donepezil, tacrine, and rivastigmine, all of which are inhibitors of acetylcholinesterase. None is a cure for Alzheimer's disease, nor do any significantly slow the progressive mental decline characteristic of this condition. Rather, for some Alzheimer's patients these agents modestly improve cognitive function by bolstering the acetylcholine system. As galantamine was known to be an inhibitor of acetylcholinesterase, it was tested as a treatment for Alzheimer's disease.

Clinical trials demonstrated the effectiveness of galantamine as a symptomatic treatment for mild to moderate Alzheimer's disease.[18-22] The chief, and most consistent, finding from this work is that galantamine significantly improves cognitive function in many of these subjects, compared with Alzheimer's patients receiving a placebo. Often the symptomatic improvement was great enough to substantially lessen the demands on caregivers. Work aimed at determining the duration of the beneficial effect of galantamine demonstrated that the drug must be given repeatedly to maintain the response.[23] This supports the notion that inhibition of acetylcholinesterase, and the resultant elevation in brain levels of acetylcholine, are primarily responsible for the clinical benefit as these effects are known to be transitory, lasting only as long as the drug is present in the brain. It also demonstrates that galantamine, like the other acetylcholinesterase inhibitors used to treat this condition, does not affect the underlying cause of the disorder.

Because some Alzheimer's patients fail to respond to galantamine, it was suspected there may be genetic differences in acetylcholinesterase that might render individuals more or less susceptible to the positive benefits of such agents. While genetic studies found that some patients metabolize galantamine more readily than others, this difference alone does not account for the variability in response.[24] Efforts are ongoing to identify a biochemical marker that can be monitored to predict the likelihood of response to galantamine and other drugs.

Laboratory experiments indicate that galantamine and galanthine display analgesic activity in animal models of pain.[25] This may explain the numbing effects associated with the application of daffodil bulb extracts for treating muscle aches, sprains, and sore joints. It is also known that an increase in pain threshold is a characteristic of compounds that enhance acetylcholine neurotransmission, including acetylcholinesterase inhibitors. It is possible, therefore, that the results of these laboratory studies reflect an effect of galantamine on pain transmission within the central nervous system by its action on the brain acetylcholine system, and a separate, local anesthetic effect, on sensory neurons. As analgesia and numbness are not usually experienced when galantamine is taken for Alzheimer's disease, it appears the doses employed clinically are insufficient to induce significant analgesia.

Both galantamine and lycorine display anti-inflammatory activity in laboratory tests.[26-28] Although attempts have been made to identify precisely the mechanism of action of this effect, it remains unknown, as does the clinical importance of this discovery.

The antitumor activity of lycorine appears to be related to an effect on cellular structure and perhaps by affecting the ability of the cell to utilize vitamin C.[29-32] It has not been established whether the concentrations of lycorine needed to reduce tumor growth can be achieved in humans. Given its reportedly brief duration of action, and its powerful emetic effect, it seems unlikely that lycorine will be a useful chemotherapeutic agent. However, continued work to characterize more fully its effects on cancer cell division, and its mechanism of action as an anticancer agent, could lead to the synthesis of chemical analogs with superior pharmacokinetic and pharmacodynamic properties.

Lycorine has also been reported to prevent the replication of human enterovirus 71.[33] As this effect is noted in an in vivo animal model, it appears the alkaloid can achieve systemic concentrations in mice sufficient to mitigate the responses to this pathogen. In humans the clinical consequences of infection include significant neurological complications. The mechanism of action of lycorine as an inhibitor of viral replication is undefined. Even though there are currently no vaccines for immunization, and no drugs to treat this infection once it has occurred, the clinical value of lycorine for this purpose may be limited by its toxicity.

The emetic effect of lycorine is thought to be due to an action in the central nervous system.[12,34] In dogs, vomiting continues for up to three hours following the subcutaneous administration of lycorine.

Data suggest this effect is the consequence of stimulation of neurokinin receptors. Neurokinin is a peptide found in brain and other tissues. A peptide is a compound composed of amino acids. A neurotransmitter, neurokinin can influence nervous system function. Lycorine is thought to induce nausea and vomiting through an interaction with neurokinin receptors because its emetic effect is blocked by administration of a neurokinin receptor antagonist.[34] Besides being present in the brain, neurokinin is known to be involved in mediating pain and inflammatory responses throughout the body. Thus, an interaction of lycorine with the neurokinin system could explain many of the pharmacological responses associated with daffodil extract.

## Adverse Effects

The side effects encountered with galantamine therapy are typical of an acetylcholinesterase inhibitor. Most common are gastrointestinal symptoms, including nausea and vomiting.[35] More troubling, particularly for the elderly, is the bradycardia, or slowing of the heart rate, that occurs as a result of activation of an acetylcholine system that innervates the heart. It has been reported that patients receiving galantamine are much more likely to withdraw from a clinical trial than those receiving a placebo.[19] This suggests that even therapeutic doses of this drug cause perceptible side effects in many subjects. While no serious safety concerns were raised after multiple, double-blind clinical studies with galantamine, in 2005 the United States government issued an alert indicating that the mortality rate was higher for patients taking galantamine than for untreated subjects.[23] A direct causal relationship between the use of galantamine and the increase in mortality has yet to be established, however.

Nausea and vomiting are the most consistent and predictable side effects associated with the consumption of lycorine. Prolonged administration can cause a scurvy-like condition, perhaps because of its ability to interfere with the utilization of vitamin C.[31]

Those handling narcissus routinely, such as florists, sometimes experience "daffodil itch." This condition is characterized by dryness and scaling of the hands, and thickening of the skin beneath the nails. This is not thought to be due to any of the alkaloids in the plant, but rather to calcium oxalate, a poisonous substance produced by many plants, including narcissus.

## Pharmacological Perspective

Daffodils, along with other members of the Amaryllidaceae family, synthesize a number of bioactive alkaloids. It is important they be separated, studied, and administered individually because while some are useful therapeutics, others are highly toxic. In fact, the presence of these toxic alkaloids has limited the systemic use of daffodil extract as an herbal supplement.

There is no question that galantamine, an alkaloid produced by daffodil and other plants, is effective in easing some of the symptoms of dementia. It has also been established that this beneficial effect is due to its ability to inhibit the metabolism of acetylcholine and to interact with certain acetylcholine receptors in the brain. Much of the pharmacological data with lycorine are also definitive, with there being little doubt that this substance is responsible for much of the nausea and vomiting that occurs following consumption of daffodil.

The high degree of certainty regarding the clinical benefits and toxicities associated with daffodil contrasts markedly with how little is known about active compounds in many other herbal supplements. Studies with daffodil illustrate the importance of identifying a pharmacologically active product for establishing the therapeutic benefits, and limitations, of a plant extract. Besides demonstrating the medicinal value of alkaloids, the studies with daffodil confirm that natural products that influence central nervous system activity can be expected to display a number of untoward side effects. This is understandable because any compound that modifies the activity of an organ as compact and complex as the brain is likely to have a number of effects on behavior, some of which will be beneficial, and some detrimental. Given this general principle, skepticism should greet any claim that a plant extract or constituent has positive effects on central nervous system function while being devoid of side effects.

# 13

## Passion Flower (Passiflora incarnata)

Unlike the many herbs that originated in Asia and the Mediterranean region, passion flower is native to the American continent. Written accounts of its use date only from the fourteenth century with the arrival of the Spanish in the Americas. As it was obvious to the Europeans that use of passion flower was a common custom among the locals, it is likely this plant had been consumed for centuries as a nutrient and herbal remedy. Today, passion flower is grown throughout the world.

The name passion flower was not bestowed on this plant because it was thought to enhance sexual attractiveness or performance. Rather, to the Spanish missionaries the components of the complex flower symbolized items associated with the passion of Christ. For example, the radial filaments represented the crown of thorns, the tendrils Roman whips, and the pointed leaf the tip of a lance.

There are hundreds of species of passion flower and a similar number of hybrids. Two that are commonly used as herbal supplements are Passiflora incarnata, which is native to the southern United

States, and Passiflora edulis, which originated in portions of South America. In Latin, *incarnata* means flesh-colored, describing the flower petals of this species, and *edulis* means edible, which refers to the fruit produced by the plant. Some of the common names for Passiflora incarnata are maypop, purple passion flower, and wild apricot. Passiflora edulis is better known as passion fruit. In Brazil, members of this genus are referred to collectively as maracuja.

Native Americans consumed passion fruit for its nutritional value and calming properties. By the eighteenth century European writers were describing the plant as a treatment for epilepsy and insomnia. Today, the herb is employed primarily to relieve anxiety and to facilitate sleep. As these uses imply a pharmacological effect on the central nervous system, attempts have been made to identify the chemical constituents of passion flower responsible for them. A review of the scientific literature suggests no consensus on the existence or identity of a pharmacologically active compound, and no definitive clinical data proving that passion flower extract is a reliable and effective anxiolytic and hypnotic.

## Botany

Passiflora incarnata is a member of the passion flower family, Passifloraceae. There are more than 500 species of passiflora scattered throughout the world. Insects and hummingbirds are responsible for pollination. Found in warm climates, passion flower thrives in South America and the West Indies. Nine species are native to North America, with Passiflora incarnata, or passion flower, and Passiflora lutea, or yellow passion flower, being among the most common. Passiflora incarnata is a perennial climbing vine that grows up to 24 feet in height. The flowers are produced in abundance from May to October. The fruit, which is referred to as maypop or passion fruit, is three inches long and ellipsoidal, much like a plum. The flesh is considered pleasant tasting and cooling. Passiflora lutea grows to about half the height of incarnata. The flowers of Passiflora lutea are light yellow, and its purple fruit is about one inch long. Passiflora edulis, the native South American species, is between 15 and 20 feet in height and produces a three-inch ovoid fruit that can vary in color between yellow and deep purple.

Some 16 species of passiflora yield edible fruit. The flowers of most species are large and sweet smelling. The plants are grown for their floral appearance and for the fruit. It is believed that Passiflora incarnata was cultivated by Native Americans for both the fruit and as an herbal remedy. Passion flower grows wild in the southern half of the United States, reaching as far north as southern Kansas and Missouri, and east from the Rocky Mountains to the Atlantic coast.

The Passiflora incarnata herb is listed in the European Pharmacopoeia. It is recommended for the treatment of anxiety and insomnia. In the United States, passion flower herbal tea is employed for the same purposes.

## Therapeutic Uses

Passion flower extract is available commercially as a dry powder, or in a capsule or liquid formulation. The herb was used by Native Americans as a systemic tonic, and to treat swellings and sore eyes.[1] Today, Passiflora incarnata extract is approved in Germany as a treatment for nervous restlessness, mild insomnia, and gastrointestinal complaints of nervous origin.[2] While this product is consumed primarily for relieving anxiety and improving sleep, laboratory animal research suggests other possible uses. These include treatment for opiate withdrawal, neuralgia, cardiac arrhythmias, and inflammatory conditions.[3-5] Extracts derived from various species of passion flower have been reported as treatments for pain, asthma, diarrhea, dysmenorrhea, burns, and hemorrhoids. This extract is also purported to be useful for lowering blood pressure, alleviating headache, and treating diabetes and skin conditions.[6]

Much of the data on the therapeutic properties of passion flower are difficult to interpret as different plant species are employed in these studies. Further complicating matters is that the portion of the plant used to produce the extract may differ among studies, as may the route of administration and dosage. As the chemical constituents in a passion flower extract differ, both qualitatively and quantitatively, among species, it is not surprising that reported biological responses to the extract vary among investigators. Still, it is unlikely that any single plant genus, regardless of the number of species and variety of chemicals produced, would display such a wide range of beneficial effects. To define more precisely the clinical utility of passion flower it is necessary to identify precisely the species being studied and, ideally, the chemical compounds in the extract responsible for any response.

## Constituents

Passion flower extract is derived from the aerial portion of the plant, with or without the flowers. The herb was listed in the United States National Formulary from 1916 to 1936. The National Formulary provided methods and criteria for pharmacists and others for standardizing herbal products in use at that time. Dried leaves, with portions of the stem, are the main plant constituents used for making the extract. Today, standard preparations of the herb contain up to 2.5% flavonoids, most as glucosyl derivatives. The major flavonoids present in these preparations are chrysin, vitexin, isovitexin, schaftoside, isoschaftoside, orientin, isoorientin, homoorientin, apigenin, and swertisin.[7,8]

Alkaloids are another chemical class represented in passion flower extracts. Individual agents include harmane, harmol, harmaline, and harmalol.[6] Maltol, a flavor enhancer found in many plants, is also produced by passion flower. Numerous other chemicals and chemical classes have been identified in passion flower extracts. Undoubtedly there are scores of other, as yet unidentified, chemicals in these products. The flavonoids and alkaloids in all species of passion flower are found in other plants. As the concentrations of the alkaloids in passion flower are much lower than for the flavonoids, the latter are thought to be most responsible for the pharmacological responses associated with this plant.[2,9]

There is considerable variation in the concentrations of the flavonoids and other compounds among passion flower species and in different portions of the plant.[10] The leaf extracts display the most anxiolytic activity in mice, followed by stem and flower extracts, with root extracts being incapable of inducing this response.[11]

Genetic diversity among the species is a major determinant of the chemicals produced by the plant. For example, two distinct chemotypes found in passion flowers are cultivated in Australia. The most common of this duo produces an extract with high concentrations of isovitexin, schaftoside, and isoschaftoside, while the other has high levels of swertisin and low levels of schaftoside and isoschaftoside.[12] Thus, plants appearing identical to the naked eye, and even upon microscopic examination, can yield different products. Furthermore, as with all flora, differences in the chemical composition of an extract, even from the same cultivar, occur with variations in the age of the plant, the time of harvest, and local growing conditions. The solvent used and the method of extraction are also important variables in determining which plant constituents and their concentrations appear in the final product. This might explain why in some, but not other, animal studies it was found that an aqueous extract was more sedating, while ethanol and methanol extracts were more anxiolytic.[11,13] Others found that a chloroform extract yielded the most active anxiolytic activity when tested in mice, while an ethanol extract was more effective than an aqueous solution in suppressing free radicals and preventing tissue damage.[4,14] Because of differences in source materials, and the resultant variability among extracts, it is virtually impossible for investigators to replicate precisely the work of others. It also poses a challenge for the consumer with regard to the consistency of commercial products.

Chrysin and vitexin are two passion flower flavonoids suspected of being pharmacologically active. Others receiving attention are orientin, apigenin, and swertismarin. Samples of these purified substances were examined in vivo and in vitro to assess the likelihood that one, or perhaps a combination of them, is responsible for the reported therapeutic effects of passion flower. Such data are the most reliable and reproducible as they are generated using a single, chemically identified, substance. The majority of published studies, however, employ a passion flower extract, limiting interpretation of the findings. Given the complex taxonomy of this genus, morphological similarities among the numerous species, the variety of chemicals present in the plant extract, and the variations in the chemical content of the plant and extract in genetics, in extraction method, and other factors, the literature on the pharmacokinetic and pharmacodynamic properties of passion flower is often contradictory, confusing, and inconclusive. If passion flower extract is affecting central nervous system function, the current evidence is unpersuasive that the chemical constituent responsible for its action has been identified.

## Pharmacokinetics

A great deal of in vitro data suggest that plant flavonoids in general can influence biochemical responses in a variety of tissues and organs, including the brain. However, only a small fraction of this compound class has been subjected to detailed pharmacokinetic studies.

There has been interest for some time in chrysin pharmacokinetics because of its purported effects on hormone metabolism and brain function. The intestinal absorption of chrysin is limited due to metabolic transformation in the gastrointestinal system and the presence of an efficient transport system that prevents significant quantities from entering the bloodstream.[15] Such findings raise doubts about chrysin being responsible for any of the pharmacological effects of passion flower extract.

In other studies, four passion flower flavonoids, isovitexin, vitexin, orientin, and homoorientin, were administered orally to rats to examine their metabolism and excretion.[16] As the majority of the administered dose of these agents was found in the feces, it appears these flavonoids are not readily absorbed into the systemic circulation. When given alone to rats, the maximum blood levels of vitexin were attained in less than an hour following oral administration, and its serum half-life was approximately two and a half hours.[17] The liver and kidney were the organs with the highest concentrations of this compound. Notably, none was found in the brain. As most of the administered dose of vitexin was recovered in the feces, it was concluded that its absorption is limited following this route of administration. Indeed, it is estimated that only 3% to 4% of an orally administered dose of vitexin makes its way from the gastrointestinal system into the bloodstream.[18] Thus, the bulk of the ingested agent never exits the digestive tract. Of the small fraction that does reach the circulation, the majority is rapidly metabolized in the liver. These findings, especially the absence from the brain, make it appear unlikely that vitexin is responsible for any of the central nervous system effects of passion flower.

With regard to apigenin, studies in rats indicate it is slowly absorbed following oral administration. However, its appearance in blood is limited by the fact that it is metabolized by intestinal bacteria and, once absorbed, rapidly metabolized by the liver.[19-21] A study with human volunteers revealed that small quantities of apigenin appear in blood following the consumption of parsley.[22] No information is available on whether apigenin appears in the brain when it is consumed orally and, if so, what the concentrations are when it is taken in this manner. Inasmuch as its absorption from the gastrointestinal system is limited, and a significant fraction is metabolized before it can be absorbed, there is little likelihood that the brain concentrations of apigenin are sufficient to induce changes in central nervous system function.

Although luteolin appeared in rat and human blood after it was taken orally, a significant portion was metabolized in the intestine and liver, limiting the amount available for mediating pharmacological effects.[23] While in rats it appears luteolin absorption occurs primarily by passive diffusion, up to one-third of the quantity ingested is excreted in the feces. It is evident, therefore, that only a limited amount of luteolin finds its way into the bloodstream even when it is administered in solvents that would be expected to facilitate its absorption.[24,25] It is likely that the passage of luteolin from the human gastrointestinal system into blood would be much less than that reported in these laboratory animal studies as the extract formulation used clinically would not be as optimal for its absorption.

As for orientin, very little, if any, of this compound was found in rat brain following its intravenous administration.[26] This indicates that orientin is not sufficiently lipid soluble to cross into the brain even if significant quantities are present in blood. Such a finding suggests that, like the other passion flower flavonoids examined, orientin is not a good candidate as the constituent responsible for central nervous system effects.

Laboratory studies demonstrate that passion flower extract, and some of the individual flavonoids contained in this product, have the potential to influence the response to other drugs. In vitro experiments revealed that this extract can interfere with a transport system that restricts the absorption of drugs used to treat breast cancer.[27] If such an interaction occurs in humans, co-administration of the extract with these chemotherapeutics could enhance their effectiveness. Conversely, these data also indicate that by affecting this efflux transporter system, the passion flower extract could facilitate the absorption of toxic agents that would normally be excreted before gaining entry into the systemic circulation.[28]

There are numerous reports on the ability of chrysin, apigenin, and luteolin to affect the activity of drug metabolizing enzymes.[29-33] Both enhancement and inhibition of enzyme activity have been reported, depending on the type of in vitro analysis performed and the particular enzyme examined. As such studies are generally conducted in vitro, their relevance to what occurs in humans consuming this supplement is unknown.

These pharmacokinetic data indicate that passion flower flavonoids penetrate poorly into the bloodstream following oral administration. For this reason alone, they are not good candidates as the chemical constituents responsible for the pharmacological effects of passion flower extract. The finding that some of these flavonoids can influence the transport of certain drugs and toxic agents into and out of the gastrointestinal system could be of clinical significance since this does not necessarily require absorption into the systemic circulation. If consumed in sufficient quantities, passion flower extract could very well modify the absorption of other substances as the herb travels down the gastrointestinal tract.

## Pharmacodynamics

The fact that passion flower extract affects central nervous system function has been demonstrated in a host of in vivo laboratory animal studies. In particular, a number of investigators have reported that administration of the extract to rats or mice induces antianxiety, or anxiolytic, and sedative effects.[9,11,14,34] What complicates these findings is that extracts from different plant species are used for these studies, different solvents are employed for the extraction procedure, and various doses are tested using different routes of administration. With regard to plant species, extracts from Passiflora incarnata, edulis, alta, and quadrangularis are just some of those examined.[14,35-37] Solvents used to prepare these extracts include water, alcohol, butanol, petroleum ether, and chloroform. The doses of extract administered ranged from less than 0.4 to up to 600 mg/kg, administered either orally or by intraperitoneal injection. While some investigators report an anxiolytic response to an aqueous extract, others found that an aqueous fraction is inactive in this regard.[35,36,38] There are also marked differences in the anxiolytic response to extracts prepared from butanol and chloroform.[14,36] Experiments with isolated flavonoids indicated that, when administered alone, both chrysin and apigenin display anxiolytic activity in rats and mice.[39-41] In these studies, chrysin was also shown to cause sedation and to induce skeletal muscle relaxation. Others, however, suggest that because the aqueous passion flower extract contains only small quantities of flavonoids, this compound class, which would include chrysin and apigenin, could not possibly be responsible for the antianxiety and sedative responses to this herb.[42] When studied as purified substances, neither the alkaloid harmane, nor maltol, induce any behavioral effects in laboratory animals.[9]

While the vast majority of these studies suggest that passion flower extract displays anxiolytic and sedative properties in rodents, there is no consensus on the optimal extraction procedure and dose needed to induce robust and consistent responses in these subjects. This raises questions about the meaning of these results in terms of their relevance to human use, and their value for identifying the pharmacologically active substance. If a single compound is responsible for the pharmacological effects, it should be most evident when it is extracted in a particular solvent. As this is not the case, either various constituents are present in passion flower that are capable of directly modifying brain function or the behavioral changes noted in the laboratory result from nonspecific effects of extract constituents on behavior.

The therapeutic utility of passion flower extract has also been suggested by placebo-controlled clinical trials.[43-49] Volunteers who consumed one cup of passion flower tea each day for a week reported a better quality of sleep than a control group that did not receive the beverage. Given the large doses that normally must be administered to laboratory animals to evoke a behavioral effect, and the pharmacokinetic properties of the passion flower constituents thought to be responsible for its central nervous system actions, it is questionable whether the small quantities consumed in a cup of tea would be sufficient to have any objective effect on sleep.

Clinical trials have also suggested that passion flower extract is an effective treatment for generalized anxiety disorder, supporting the idea that this herb has anxiolytic effects.[45,46] A similar conclusion was drawn from other studies comparing the antianxiety response to passion flower to that associated with oxazepam, an established anxiolytic drug. Moreover, it was reported that preoperative patients who took a passion flower product had lower anxiety scores than preoperative patients not given the herbal supplement.

The clinical data on passion flower have been used to justify its listing as one of 9 plants out of 1,000 examined that displays evidence of inducing a therapeutic effect.[49] Nonetheless, passion flower clinical trials conducted thus far are limited in number and have yet to be replicated with larger patient populations and rigorous measures of efficacy.[50] While the animal and clinical experiments tend to support the use of passion flower extract as a calming agent, further studies are needed to determine whether this is a pharmacological response to the herb or a manifestation of a psychological effect. The latter interpretation seems likely inasmuch as pharmacokinetic studies have yet to demonstrate conclusively that passion flower contains any compound, or group of compounds, that penetrates into the brain following oral administration of the extract. For this reason, questions remain about the pharmacological value of this herb, the clinical and laboratory data notwithstanding.

Laboratory animal work also suggests that passion flower extract displays anticonvulsant, antidepressant, and analgesic activities.[51-55] Because passion flower extract was found to reduce the amount of St. John's wort (see Chapter 6) needed to induce an antidepressant response in animal models, it was suggested that combined treatment with these two herbal supplements may yield a superior clinical response than can be achieved with either alone.[52]

With regard to its anticonvulsant activity, in one study a Passiflora incarnata ethanol extract was shown to be effective in mice at a dose of 0.4 mg/kg, whereas in another the effective anticonvulsant dose of the same type of ethanol extract was between 150 and 600 mg/kg in mice. The large difference in doses with these ethanol extracts from the same plant species illustrates why many find it difficult to draw firm conclusions from such work. Such variations in findings do not normally occur if investigators use identical experimental protocols with a single, purified substance. They also would not be expected if passion flower produced significant quantities of a pharmacologically active agent that affects brain function.

It has been reported that orientin, a passion flower flavonoid, displays analgesic activity, whereas vitexin, another flavonoid, is inactive in conventional animal models of pain.[55,56] It is doubtful this finding has any clinical significance because neither orientin nor vitrexin readily accumulates in the brain following oral administration, and analgesia is not an effect widely reported by humans during the several hundred years this herb has been consumed.

Studies aimed at defining a possible mechanism of action for the central nervous system effects of passion flower have focused on the brain GABA neurotransmitter system. Inasmuch as GABA is an inhibitory transmitter, agents that enhance its activity are known to relieve anxiety, induce sleep, and block seizures. Because all of these responses are reported for passion flower, an effect on GABA would provide a plausible explanation for its central nervous system actions. In fact, in vitro studies with passion flower extract indicate it affects components of the brain GABA system.[3,51,57,58] The extract is reported to inhibit GABA accumulation into brain cells, which would prolong the action of this transmitter, to interact with GABA binding sites, and to induce GABA-like effects in rat brain slices. In vivo work in laboratory animals has also supported a possible interaction between passion flower herb and the brain GABA system.[34] These in vivo and in vitro findings suggest that the extract contains a substance capable of directly activating brain GABA receptors. In fact, it is now known that the active constituent is GABA itself. As is true for many plants, passion flower, like mammalian brain, produces GABA. This means that GABA is present in the plant extract.[51,59] Because of this, the in vitro results indicating that passion flower extract activates GABA receptors, interacts with GABA binding sites, and inhibits GABA accumulation in the brain can all be easily explained by the presence of GABA in the test sample. As GABA is not absorbed following oral administration, nor does it readily cross into the brain, the GABA in the passion flower extract cannot be responsible for any of the reported behavioral effects of this herb. Accordingly, the significance of the data suggesting that passion flower extract interacts with the brain GABA system is suspect unless the investigators demonstrate they have removed this amino acid from the experimental sample. As this is not routinely done, no definitive data exist indicating that passion flower extract selectively interacts with GABA or any other neurotransmitter system in the brain.

With regard to effects outside the central nervous system, there was for some time the belief that chrysin decreased estrogen levels. This idea resulted from the discovery that, in vitro, chrysin inhibits aromatase, an enzyme responsible for the conversion of testosterone to estrogens.[60] If this occurred in vivo, such an effect could be exploited for treating some types of cancer and for increasing testosterone levels. This latter effect made chrysin-containing foods popular among men who thought such an action could increase their strength, energy, and sex drive. After years of study it was concluded that while chrysin is capable of inhibiting aromatase in vitro, it is inactive in vivo, most likely because of its poor bioavailability and rapid metabolism.[61-64] This illustrates the importance of pharmacokinetics in determining the responses to drugs and natural products. It also demonstrates further that chrysin is an unlikely candidate as the passion flower component responsible for central nervous system effects.

Other responses to passion flower and its constituents that have been reported are a reduction in free radicals and an anti-inflammatory effect.[4,65,66] Both of these actions are characteristics of many flavonoids. These effects are used to explain laboratory animal findings that passion flower extracts protect heart muscle from ischemic damage, display antiulcer activity, and are effective as treatments for inflammatory bowel disorder, allergies, and major depression.[67-71] There are also suggestions that passion flower extract, or individual passion flower flavonoids, have antidiabetic and anticancer effects.[72,73] As no definitive clinical studies support any of these uses, and it is known that the flavonoids shown to be active in vitro are poorly absorbed in humans, it seems unlikely that passion flower extract, or components thereof, are of any clinical value in the treatment of these conditions.

## Adverse Effects

While in one clinical study it was noted that the administration of passion flower caused drowsiness, dizziness, and confusion in some subjects, the herb appears to be generally free of side effects when taken alone at the recommended dose.[45] Given its potential to reduce central nervous system activity, caution should be exercised when taking passion flower in combination with drugs and other agents, such as alcohol, that are known to be sedating. This is illustrated by the report of an individual who experienced dizziness, tremor, and feelings of fatigue after consuming passion flower with valerian (see Chapter 7) and lorazepam, both of which are known to decrease central nervous system activity.[74]

Passion flower extract, and some of its individual components, can modify the absorption and excretion of drugs. Such effects are due to the interactions of the passion flower constituents with transport systems in the intestines and other tissues, and with their ability to modify the activity of liver enzymes responsible for drug metabolism. Although questions remain about the extent to which passion flower constituents enter the bloodstream after oral administration, and therefore the likelihood they will affect these drug transporters and metabolizing enzymes, the possibility exists that consumption of this herbal supplement, especially over extended periods, could influence responses to prescription medications that are taken at the same time.

## Pharmacological Perspective

It is a fact that many individuals find consuming passion flower extract relieves tension and anxiety, and facilitates sleep. To deny this it would be necessary to ignore or discount the hundreds of years this herb was used for these purposes, both in prehistoric times and in the modern era. Moreover, there are ample laboratory animal and clinical data demonstrating these effects. What is uncertain is whether these actions represent a pharmacological response to chemicals contained in the passion flower extract or are, in humans at least, nonspecific behavioral effects that occur only in those who believe this herb has such beneficial actions. Behavioral studies on placebo responses demonstrate that humans are able to experience all types of emotions and to obtain symptomatic relief from numerous conditions, simply by believing, even if incorrectly, that they have consumed an effective medication. Such a placebo response could also explain the popularity of passion fruit among Native Americans. That is, cultural influences and expectations can have a significant effect on the behavioral response to any substance, whether or not it has a direct effect on the brain. The degree to which the responses to passion flower are psychologically, rather than pharmacologically, mediated remains unresolved given the experimental inconsistencies and contradictory findings with this substance, and the lack of success in identifying a chemical constituent responsible for these effects.

Some individuals are persuaded of the pharmacological actions of passion flower by the results from in vivo laboratory animal studies. After all, rats and mice are not likely to be psychologically primed to respond to this herb in a certain manner. However, there are many reasons such animal data cannot be taken as proof that a compound or plant extract has a particular effect in humans. Among these is the lack of information on the relative difference in the doses used in the laboratory as compared to those taken by humans. Without precise knowledge about the pharmacokinetic properties of the active constituent, it is impossible to know how much a human would have to swallow to consume an amount equivalent to what the laboratory animal receives when it is administered 600 mg/kg of the extract. While it is possible that some passion flower substances enter the brain and induce a behavioral response in rats when administered at this dose, such an effect would never occur clinically if the animal dose far exceeds the amount recommended for human use.

Administration of passion flower extract does have a calming effect for some individuals, although the scientific evidence supporting a pharmacological basis for this response is weak. The lack of solid laboratory and clinical data will not, and should not necessarily, deter the use of this product for those who find it beneficial. Nonetheless, many will remain skeptical about the value of this extract for any therapeutic purpose until one of its constituents is shown to exhibit the pharmacokinetic and pharmacodynamic properties expected of a substance, or of a defined mixture of plant constituents, that would be capable of altering behavior.

# 14

## Coffee, Tea, and Cocoa

Caffeine, a central nervous system stimulant, is the most widely used plant product in the world. Virtually everyone has experienced its pharmacological effects. It is estimated that the amount of caffeine consumed daily is the equivalent of one cup of coffee for every man, woman, and child on Earth. This rate of intake is understandable as caffeine is a component of numerous consumer products. Besides coffee, caffeine is present in tea, cocoa, chocolates, maté, soft drinks, and various over-the-counter medications. The chief plant sources for caffeine are the shrubs Coffea arabica and Coffea canephora, and the trees Camellia sinensis and Theobroma cacao. The species epithet arabica was given to the coffee plant because of its place of origin. *Camellia* is the latinized version of Kamel, the surname of a Jesuit pharmacist and botanist who lived in the Philippines, and *sinensis* is Latin for "from China." *Theobroma* is Greek for "food of angels," and *Cacao* derives from the Aztec term *cacahuatl*, which means "bitter water." This describes the taste of the brew made from the unprocessed chocolate. Cacao is the name of the plant and its extract, whereas cocoa is the beverage made from the chocolate produced by this plant.

The genus Coffea is native to Africa and the Arabian Peninsula, Camellia to China and Southeast Asia, and Theobroma to tropical portions of the Americas. Over the centuries, armed conquest, exploration, and trade routes spread these plants, and their uses, around the globe. The popularity of caffeine has continued to grow unabated. Although it is possible to synthesize the caffeine molecule, it is still chiefly obtained from plants.

Prior to the fifteenth century, coffee beans were produced only in Africa. The use of coffee as a beverage spread from Ethiopia and Yemen to Egypt and then to Turkey. Bunchum, an Arabian coffee, was first described in the ninth century by al-Razi, a Persian physician. It was noted that consumption of coffee could prolong the number of hours worked. By the seventeenth century coffee was being enjoyed throughout Europe.

The documented history of tea predates that for coffee. Records indicate that tea was prepared for Chinese royalty as early as 3000 BC.

Archaeological findings suggest that beverages from the cacao plant were consumed in Mexico 4,000 years ago. Constituents of the cacao bean have also been identified in 2,600-year-old Mayan pottery. It is known that Mayans held an annual celebration to pay homage to the cacao god. It was used by Native Americans as an herbal remedy and for cooking. Their awareness of its stimulant properties is evident from writings indicating they consumed a chocolate drink to relieve fatigue. It is thought that Columbus was the first European to taste a cacao bean. Cacao and chocolate remained unknown in Europe until the seventeenth century when the plant was brought to the European continent by Spanish explorers.

Caffeine is an excellent example of a plant product that can affect brain function. Because it is derived from a variety of unrelated species, no single plant is highlighted in this chapter. Rather, the focus is to consider the sources and pharmacological properties of caffeine to demonstrate how such information is useful for defining the clinical utility of a plant constituent. A discussion of caffeine also emphasizes the pharmacological potential of alkaloids as compared with other types of plant chemicals. Whereas the relatively recent discovery of galantamine in daffodil (see Chapter 12) illustrates how difficult it can be to identify plant constituents having subtle effects on brain function, caffeine is an example of how straightforward the discovery process is when the central nervous system effects are obvious. This provides an important lesson for consumers interested in assessing the potential effectiveness of an herb for treating a central nervous system disorder. As shown by caffeine, it is relatively easy to obtain data demonstrating a definitive, predictable, and consistent response to a plant constituent if its effects are readily apparent. While the stimulating effects of these plants have been appreciated for thousands of years, it was not until the nineteenth century that caffeine was first isolated from them. This made it possible to demonstrate that caffeine was the constituent most responsible for the central nervous system effects of these plants. Unlike many of the herbal products sold today, it was evident to all even before the modern era that these plants contain pharmacologically active constituents.

## Botany

The plants used for preparing the more popular caffeinated beverages are botanically diverse. The most common tea plant is one of the 100 species of the genus Camellia in the family Theaceae. The camellia flower comes from a different species, Camellia japonica, which is native to Japan. Wild varieties of the tea plant are found mainly in southwestern China, South and Southeast Asia, and eastern India. The flowers are yellow to white. Tea is prepared from the leaves. Virtually all varieties of tea are produced from Camellia sinensis. This includes white, green, and black teas, as well as oolong. Their different tastes result from the way they are processed, which causes variations in the concentration of chemical constituents. Inasmuch as this species has been cultivated for millennia, it is no surprise that many different cultivars have been developed. Camellia sinensis is a perennial, evergreen tree. Cultivated varieties are usually kept at two to six feet in height. In the wild they can be up to 100 feet tall.

Leaves of Ilex paraguariensis are used to make maté, a South American tea. Ilex is a member of the holly family, Aquifoliaceae. It grows wild in warm regions of South America. Other members of this family are found throughout Europe, Asia, and the Americas. One species, Ilex vomitoria, is native to the southern United States. Native Americans made a tea from its leaves.

Coffee beans are the seeds of Coffea arabica, a shrub that grows wild in the tropical areas of Ethiopia and the Sudan. Caffea is a member of the Rubiaceae family, a varied and cosmopolitan group of mostly tropical, both woody and herbaceous, plants. Coffee trees grow up to 40 feet in height. The leaves are a glossy green and the flowers white. The fruit, which is bright red, is composed of an outer fleshy layer that protects the pit which, normally, contains two seeds. It is these seeds that are referred to as coffee beans.

Theobroma cacao is the taxonomic designation for the evergreen that produces cocoa beans, the source of chocolate. Caffeine is a constituent of chocolate. Theobroma belongs to the family Sterculiaceae, of which there are 22 species. A tropical plant native to the Americas, the cacao tree is about 25 feet tall at maturity. It produces small flowers that are pollinated by flies, rather than bees or butterflies. The seeds, or beans, from which cocoa powder and chocolate are prepared are contained in its fruit, or pod, which weighs nearly a pound. It is believed the plant originated in the Amazon region of South America. Products from other members of this group include kola nuts from Cola acuminata, which is found in tropical America, and nuts produced by various species of Sterculia in tropical Asia.

The common feature of all of these plants is that they produce caffeine, a pharmacologically active substance that affects various organ systems, including the brain. These effects popularized their use among native peoples from diverse and geographically distinct cultures.

## Therapeutic Uses

Historically, caffeine-containing plants were consumed because of their ability to increase alertness and to diminish fatigue. These effects are now known to be due to the presence of three alkaloids in the plant extracts: caffeine, theophylline, and theobromine. As a group, these compounds are known as methylxanthine alkaloids.

It was noted in a sixteenth century text that the Chinese drank tea because they found the beverage invigorating and felt that it improved their mental abilities.[1] In China the leaf of Camellia japonica, the species most noted for its flowers, was mixed with oil and applied topically to treat burns. Leaves of Camellia sinensis were used as a poultice for treating skin ulcers, swellings, and bruises. A strong decoction of tea was also used for washing wounds and sores.[1] In addition, the Chinese used tea for treating asthma and cardiovascular disorders. In India, a beverage prepared from Camellia sinensis was applied to inflamed gums.[2]

There are few written records on the early medicinal use of chocolate. Nonetheless, it has been noted that mixtures containing cacao were employed by Native Americans for therapeutic, as well as ritualistic and culinary purposes. Since its isolation and purification in the late nineteenth century, theobromine, the main alkaloid in cacao, has been taken to enhance blood circulation, to stimulate the heart, and as a diuretic to increase urination. Given these effects, the conditions that were once treated with this alkaloid include atherosclerosis, angina, and high blood pressure. Recently it has been suggested that theobromine may be a useful cough suppressant.[3]

In Western medicine, caffeine and caffeinated beverages were in the past employed clinically to stimulate respiration and as diuretics. The latter effect was beneficial in reducing the fluid accumulation associated with various medical conditions, such as heart failure. Caffeine has also been prescribed to treat apnea, a temporary cessation of breathing, in infants. Theophylline has been taken on its own as a diuretic, for relieving the symptoms of asthma, and for treating symptoms associated with chronic obstructive pulmonary disease (COPD). Because consumption of methylxanthine alkaloids stimulates cardiac muscle, they have been administered to increase heart rate and blood flow. Theobromine, in particular, has been utilized for this purpose.

A number of human studies have been aimed at determining whether there is a beneficial correlation between the consumption of caffeinated beverages and certain medical conditions. There are reports that the daily consumption of two or more cups of caffeinated, but not decaffeinated, coffee reduces the incidence of depression.[4] In a separate, smaller study, it was suggested the incidence of depression was reduced if polyphenols were added to decaffeinated coffee.[5] There have also been reports that coffee consumption may be linked to a decrease in the risk for breast and colon cancers, for type 2 diabetes, and for cardiovascular disease.[6-10] While such findings are interesting, especially for the millions who consume significant quantities of coffee, such correlations do not prove a causal link between caffeinated drinks and the reported protection from these conditions. Further studies are needed, preferably with single, purified compounds, to demonstrate these clinical benefits conclusively.

Caffeine-containing consumer products are still marketed as aids for staying alert and awake. In the past, caffeine has also been a component of drug combinations to counter the sedative effects of anti-inflammatory and analgesic agents. Today, however, none of these methylxanthine alkaloids are used routinely for therapeutic purposes. Rather, for the most part, they are consumed as ingredients of coffees, teas, chocolate, and soft drinks.

Beyond the alkaloids, there has been a growing interest in the polyphenol content of caffeinated beverages, especially teas prepared from Camellia sinensis. Data suggest that these plant products may have beneficial antioxidant and anti-inflammatory activities. It has also been proposed that polyphenol dietary supplements may help induce weight loss.[11]

## Constituents

Scores of plant chemicals are contained in beverages brewed from coffee beans, cacao beans, or tea leaves. Of these, the most characterized, and the ones that have thus far demonstrated the most pharmacological activity, are the alkaloids caffeine, theophylline, and theobromine. Caffeine is produced in plants through a multistep process. Theobromine is one of the chemical intermediates in the plant synthesis of caffeine. Both theophylline and theobromine, besides being present in these plants, are also produced in humans following ingestion of caffeine. Thus, the blood concentrations of theophylline and theobromine are dependent, in part, on the amount of caffeine consumed and metabolized.

Plants do not synthesize these alkaloids for the enjoyment of humans. Rather, these chemicals are insecticides that protect the plant from natural predators. It has been proposed that polyphenols, including the various flavonoids present in these extracts, may have important therapeutic effects. However, because their purported actions are more subtle than those induced by the alkaloids, the actual therapeutic benefits of the polyphenols remain a matter of debate.

Structurally, caffeine is a methylxanthine alkaloid (see Figure 14.1), as are theophylline and theobromine. The structural difference among these compounds is the location of the methyl groups on the xanthine molecule. Of these three compounds, caffeine is the most potent central nervous system stimulant.

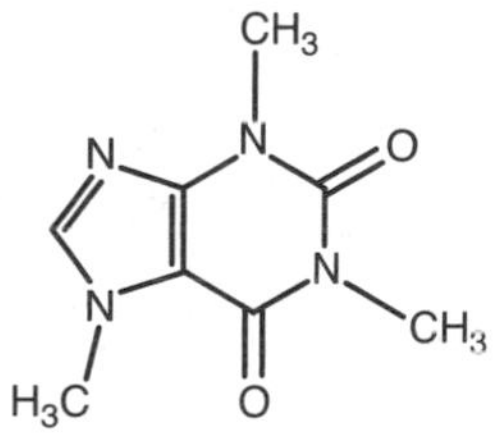

Figure 14.1 Chemical structure of caffeine (Wikipedia)

The term *caffeine* is derived from *café*, the French word for coffee. A typical cup of coffee contains approximately 100 mg of caffeine and maté about 85 mg. There is only 50 mg or so of caffeine in a cup of tea, and around 5 mg in a cup of cocoa. Theophylline is present in only trace amounts in coffee and cocoa, but in greater quantities than caffeine in teas. While there is very little theobromine in tea and coffee, up to 250 mg of this alkaloid are present in a cup of chocolate and some 40 mg in a cup of maté. Thus, coffee should be drunk if caffeine is being sought, tea for theophylline, and cocoa or maté if consumption of theobromine is the objective. Because cacao contains much lower amounts of caffeine per unit of weight, chocolate and other cacao containing products have much less stimulant effects on the central nervous system compared with coffee and tea. These distinctions are not absolute, however, as the relative amounts of the different alkaloids will vary in a beverage depending on the cultivar used and the way the bean or leaf is processed and brewed. For example, darker roast coffees generally have a lower caffeine content than lighter roasts, even though the darker may taste more bitter. The situation is the same for teas, with lighter brews, such as Japanese green tea, having a much higher caffeine content than darker brews.[12] The longer brewing process used to prepare the darker coffees and teas diminishes the caffeine content.

The flavonoid catechins are the most abundant polyphenols in tea (see Figure 14.2). The most common catechin in tea is epigallocatechin-3-gallate. As with other plant constituents, the catechin content differs widely among teas.[13] The main polyphenols in coffee are chlorogenic acid, caffeic acid, and trigonelline.[14] Polyphenols make up 10% of the dry weight of cacao beans.[15] A major polyphenol in cacao is clovamide.[16] All cacao beverages contain catechins.[17]

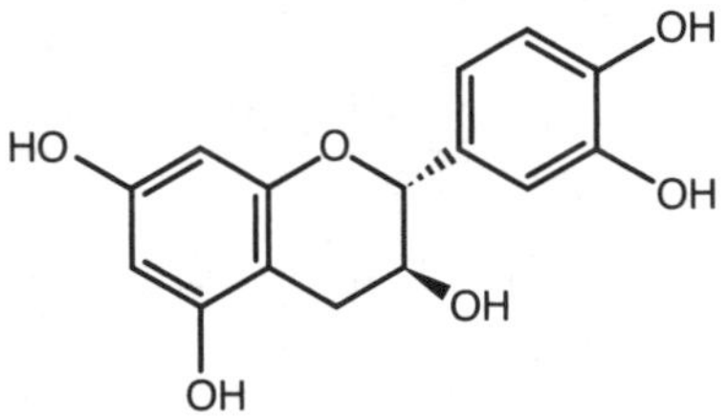

Figure 14.2 Basic chemical structure of a catechin (Wikipedia)

Maté, a South American tea, more resembles coffee than other teas in that chlorogenic and caffeic acids are the major polyphenols contained in this drink.[18,19]

## Pharmacokinetics

In both laboratory animals and humans, caffeine, theophylline, and theobromine are readily absorbed into the bloodstream after oral administration. When rats are administered about 8 mg of caffeine by mouth, the blood levels achieved are sufficient to cause behavioral changes characteristic of central nervous system stimulation.[20] In humans, the maximal plasma concentration of caffeine is attained within an hour of ingestion. The half-life for caffeine in human blood ranges from three to five hours, whereas the half-lives for theophylline and theobromine are six to seven hours.[21] The majority of a dose of caffeine, theophylline, or theobromine is metabolized in liver, with only small fractions of the parent compounds excreted unchanged in the urine.[22,23] The blood levels of caffeine following the oral administration of 100 mg, roughly the amount consumed in a cup of coffee, are sufficient to increase wakefulness.[24] The human ingestion of 1 gram or more of caffeine causes marked central nervous system stimulation, which manifests as insomnia, restlessness, excitability, and, possibly, seizures.[22]

Unlike the alkaloids, the polyphenols, such as the catechins in green tea, are poorly absorbed following oral administration.[11] Nonetheless, it is believed the consumption of these substances causes a reduction in levels of oxidative stress, a process thought to be responsible for a number of medical conditions.[25] Given their low lipid solubility, it is unlikely that any of the central nervous system's effects of coffee or tea are due to an action of polyphenols in the brain. Rather, if the catechins are pharmacologically active, their effects are probably restricted to tissues and organs located outside of the central nervous system.

Because the methylxanthine alkaloids are so dependent on liver metabolism for their elimination, any changes in the activity of the relevant metabolizing enzymes will shorten or prolong their half-lives in blood, and therefore affect the pharmacological responses to these agents. For example, theophylline is more readily eliminated, and therefore less active at a given dose, in children and smokers, whereas the response to it is enhanced in individuals with liver disease as they are less able to metabolize it efficiently. Also, drugs and other agents that are broken down by these same liver enzymes could, especially if administered for prolonged periods of time, affect the rate of elimination of the methylxanthines. Certain antibiotics, for example, are known to inhibit a liver enzyme that is important for the metabolism of caffeine, slowing significantly its destruction and thereby prolonging its action.

The pharmacokinetic data for methylxanthines produced by coffee, tea, and cacao plants indicate these agents have characteristics that are typical for drugs known to penetrate into the brain and evoke central nervous system effects. These results are strong evidence that the brain stimulation associated with the consumption of these plant products is due to these compounds. In contrast, the more limited absorption of polyphenols following oral administration casts significant doubt on their ability to penetrate into the brain.

## Pharmacodynamics

The most prominent response to caffeinated beverages is central nervous system stimulation. Evidence that this action is due primarily to the caffeine content of these plant products is provided, in part, by the finding that consumption of decaffeinated coffee has less of an effect on mood and cognitive performance than caffeinated coffee.[25] At conventional doses the caffeine-mediated brain stimulation is characterized by an increase in alertness, as well as improvements in muscle coordination and in the ability to concentrate. In both laboratory animals and humans, caffeine, theophylline, and theobromine also have direct effects on organs and tissues outside of the central nervous system. These include stimulation of the heart rate and force of contraction, dilation of blood vessels and lung bronchioles, and an effect on the kidney to increase urination. The direct action on the heart increases cardiac output, which is a measure of the amount of blood pumped from the heart with each contraction. Tolerance, or a diminished response, to these and other effects of the methylxanthines, as well as physical and psychological dependence, develop with their continuous administration. With the development of tolerance, the response to a fixed dose is diminished. Accordingly, tolerant individuals must take more of these compounds than naïve users to achieve the same effect.

Studies, primarily in vitro, suggest that polyphenols found in coffee, tea, and cocoa may have anti-inflammatory effects, inhibit the formation and growth of tumors, help prevent cardiac disease, and display antibiotic activity.[26,27] While there is little doubt these compounds display such activities in test tube assays, consistent and definitive in vivo data are lacking to support these findings.

Although caffeinated beverages have been consumed for thousands of years, it was not until the late twentieth century that the mechanism of action of the methylxanthines was defined. At the biochemical level, two prominent effects are inhibition of phosphodiesterase, an enzyme responsible for the destruction of an important intracellular chemical, and inhibition of the receptor for adenosine, a neurotransmitter substance.[24] It is believed that the reduction in adenosine transmission is primarily responsible for the central nervous system stimulant effect of the methylxanthines, caffeine in particular.[28]

Adenosine is a cellular chemical located in tissues throughout the body, including the brain. Besides its important role as a component of the energy generating system in cells, adenosine also serves as a neurotransmitter. There are several molecularly distinct adenosine receptors in the central nervous system. Two of these, the adenosine $A_1$ and $A_{2A}$ receptors, appear to be the ones most readily inhibited by caffeine. The adenosine $A_1$ receptors are highly localized in the brain stem and certain areas of the cerebral cortex, whereas the $A_{2A}$ sites are concentrated in the hypothalamus. As adenosine is an inhibitory neurotransmitter, activation of its receptor decreases neuronal activity in these and other brain regions.[29-33] It is inhibition of the action of adenosine by caffeine and the other methylxanthines that is responsible for the enhanced alertness and cognition, and the reduction in fatigue, that is associated with the consumption of coffee, tea, and cacao. It is suspected that changes in adenosine receptors play a crucial role in the development of tolerance to caffeine, and in the withdrawal symptoms experienced when the intake of this methylxanthine is halted after a prolonged period of consumption. Inhibition of adenosine receptors in the corpus striatum is thought to be responsible for the effects of caffeine on locomotor activity and muscle coordination.[28,34]

Oxidative stress in neurons may be a major factor in the cell death associated with neurodegenerative disorders, such as Alzheimer's and Parkinson's diseases, and in the normal cognitive decline experienced with aging.[35] It has been shown in vitro that some catechins present in tea, especially epigallocatechin-3-gallate, are capable of neutralizing the chemical compounds, referred to as oxygen radicals, that mediate the tissue destruction associated with oxidative stress.[36] This has led to proposals that this polyphenol may protect the brain from the ravages of oxygen radicals. There remain significant doubts about this theory because the concentrations of epigallocatechin-3-gallate needed to block the effects of oxidative stress are far above those that would be anticipated in the brain following consumption of typical amounts of tea. Moreover, a clinical trial found no evidence supporting the theory that epigallocatechin-3-gallate increases the antioxidant activity in blood.[37] To observe antioxidant activity in mice, green tea had to be infused into the bloodstream.[38] These findings are consistent with the pharmacokinetic studies suggesting that polyphenols, such as the catechins, are not well-absorbed into the body following oral consumption, making it unlikely that significant quantities of these substances find their way into the brain.

Although polyphenols may not protect brain cells from oxidative stress, their ability to reduce the production of oxygen radicals might have beneficial effects on tissues outside the central nervous system. These could include anti-inflammatory actions and inhibition of tumor cell growth.[39] However, a comprehensive review of studies aimed at evaluating whether green tea prevents cancer indicates the evidence at this time is insufficient for drawing this conclusion.[40]

## Adverse Effects

The side effects associated with the consumption of caffeinated beverages are predictable as they are extensions of their pharmacological actions. Thus, adverse central nervous system effects that may be experienced by sensitive individuals, or by anyone following the consumption of large quantities of coffee or tea, include insomnia, restlessness, and anxiety. In extreme cases, mania, disorientation, hallucinations, and seizures may occur. Other side effects include increased urination, gastrointestinal distress, muscle twitches, and tachycardia, or rapid heartbeat. This effect on heart rate could prove fatal for someone with a preexisting cardiovascular condition.

Chronic consumption of caffeine-containing beverages, especially coffee, can result in tolerance and physical dependence. Withdrawal symptoms are experienced if caffeine ingestion is terminated abruptly. These symptoms, which are relieved by coffee or administration of caffeine alone, include headache, fatigue, and drowsiness.[28]

There is little evidence suggesting that consumption of the polyphenols contained in caffeinated beverages has any adverse consequences. This is likely related to the fact that these substances are poorly absorbed from the gastrointestinal tract. Nevertheless, studies with catechins suggest that, at sufficiently high concentrations, these agents cause, rather than inhibit, oxidative stress.[41] This has been noted in vivo, with high oral doses of epigallocatechin-3-gallate producing oxidative stress-induced liver damage in mice.[42] Because of the doses needed to observe this toxic response, and the fact that tissue damage has not been reported for caffeinated beverages during their thousands of years of use, this response to polyphenols does not appear to occur in humans at the doses normally consumed in tea, coffee, and cocoa.[40]

The side effects associated with the methylxanthines provide further evidence that they can affect central nervous system function. Conversely, the lack of reported side effects for the polyphenols indicates these agents are inactive at the doses usually consumed. In part this is because their pharmacokinetic properties make it impossible for them to attain sufficient concentrations in tissues to induce a pharmacologic or toxic response. Comparison of the adverse effects of the methylxanthines and the polyphenols demonstrates again the principle that plant products with definitive actions on central nervous system function generally display side effects. The lack of adverse effects for any plant extract or other type of herbal supplement makes it unlikely that they influence brain function.

## Pharmacological Perspective

The discovery of caffeine and related methylxanthines as pharmacologically active plant products progressed in a logical and systematic manner. It was known that humans had for thousands of years consumed these plants because they found them to be invigorating and to reduce fatigue. The likelihood that these actions were mediated by a plant constituent was enhanced by the fact that the positive central nervous system effects of this diverse group of plants were discovered independently by different cultures scattered around the globe. This suggests the reported actions were not simply a culturally related psychological response or a matter of local custom. Such history, and the fact that the beneficial effects of these brewed beverages were instantly appreciated when they were first introduced in other parts of the world, provided ample justification for undertaking the work necessary to identify the active products in these plants.

Because the agents are so potent and have appropriate pharmacokinetic properties, it was relatively easy to determine in animal studies which of the many plant chemicals were responsible for the effects on the central nervous system. Once the active ingredients were identified, they were used clinically as treatments for central nervous system and cardiovascular disorders. At the same time, pharmacologists conducted studies to define their mechanism of action. When this became known, efforts were undertaken, and continue to this day, to synthesize related compounds with the aim of developing safer and more selective drugs. This sequence is an excellent example of how information obtained from scientists in different disciplines, from anthropology, to botany, chemistry, and pharmacology, contributes to drug discovery. It also illustrates that plants having significant central nervous system effects are normally easy to identify, both by primitive humans and modern scientists.

Accordingly, from a pharmacological standpoint, it is difficult to be enthusiastic about herbal supplements that are purported to affect brain function but are known to be variable in their effects and for which there is no definitive proof that they contain constituents capable of affecting the central nervous system. The discovery of caffeine, theophylline, and theobromine demonstrates that plants are a valuable resource for clinically useful drugs. The story of their discovery and use underscores the need to exercise caution when attempting to discern whether an herbal supplement has beneficial effects when the active constituent has not been positively identified or, as in the case of polyphenols, cannot be shown to reach the target tissue in the body.

# 15

## Epilogue

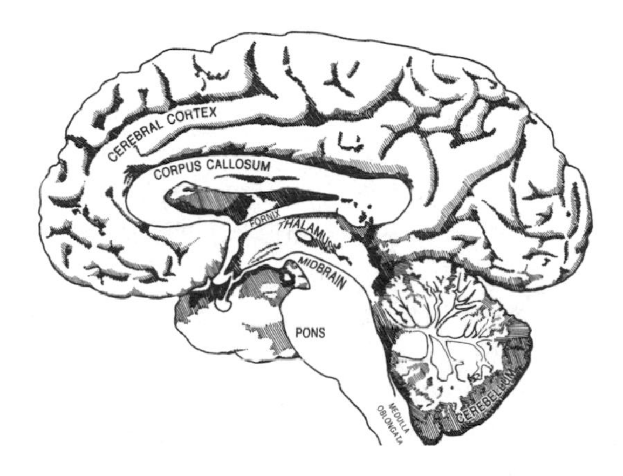


As detailed in the preceding chapters, the scientific evidence supporting the central nervous system actions of herbal supplements varies from strong, to weak, to nonexistent. Such information is crucial for making informed decisions about the use of such agents. Given the purpose of this book, an effort was made to select a range of plant materials that would demonstrate that claims about the medicinal value of these products are sometimes sound and sometimes unsubstantiated. The risk with this approach is that it implies that plant constituents are an uncertain source for such agents. This is certainly not the case. If the objective was to describe only plant products for which there are definitive data supporting an effect on brain function, many other plants and their constituents would have been included. Among these are four that should be mentioned because of their contemporary use and their value in illustrating further some of the basic principles espoused in this volume (see Table 15.1).

**Table 15.1** Some Popular Natural Products with Central Nervous System Effects

| Plant Name Common/Botanical | Active Ingredient | Effect on Brain |
|---|---|---|
| Opium poppy/Papaver somniferum | Morphine | Analgesia/Sedation |
| Ma Huang/Ephedra sinica | Ephedrine | Stimulation |
| Coca/Erythrylon coca | Cocaine | Stimulation/Euphoria |
| Marijuana/Cannabis sativa | Cannabinoids | Euphoria |

Morphine is obtained from the unripe seed capsule of the opium poppy, Papaver somniferum. Morphine constitutes 4% to up to 20% of the gum resin in the capsule. As with all plant constituents, this variation is due to differences in cultivars, climate, and horticultural practice. The responses to morphine are the result of its interaction with opioid receptors in the brain and other organs. Even after years of research aimed at creating or discovering superior analgesics, morphine remains the most effective drug for alleviating pain. Morphine and its chemical derivatives are still widely employed as analgesics. Their use is restricted to treating certain types of pain in clinical settings because of their side effects and the fact that they are addicting. The addiction liability supports the notion that morphine acts within the central nervous system. Research continues on the development of drugs that display the same range and level of analgesia as morphine, but that do not interact with the brain areas responsible for addictive behavior. Chemical derivatives of morphine include heroin, a highly addictive euphoriant which is not produced by the plant. Codeine, another chemical analog of morphine, is present in the opium poppy capsule. Codeine has for generations been used as a cough suppresant.[1] The continued use of morphine and related compounds indicates that, when it comes to drug discovery, it can sometimes be difficult to improve upon nature.

Ephedrine, the active ingredient in Ma Huang, is found in several species of Ephedra and in some unrelated plants such as Catha edulis. The content of ephedrine varies as a function of the genus and species or origin. The most common natural sources for this chemical are two Chinese plants, Ephedra sinica, or Ma Huang, and Ephedra equisetina. Ephedrine causes the neuronal release of norepinephrine, resulting in activation of the receptors for this transmitter.

A mild central nervous system stimulant, ephedrine has been taken, sometimes in combination with caffeine and aspirin, to decrease appetite and reduce weight. In the United States ephedrine is no longer approved for this purpose because of toxicities associated with its use.[2] Historically, Ephedra plants, or extracts taken from them, were used as treatments for asthma, bronchitis, and nasal congestion. It is known now that the release of norepinephrine from nerves in the respiratory system relieves symptoms of these conditions.[3] The synthesis of amphetamine, a powerful central nervous system stimulant, was based on the chemical structure of ephedrine. As compared with ephedrine, the greater central nervous system effectiveness of amphetamine is due in large measure to its superior pharmacokinetic properties, which allow for a more rapid and extensive accumulation in the brain. The development of amphetamine is an example of science improving on nature if the objective is to make a superior central nervous system stimulant.

A local anesthetic and drug of abuse, cocaine is obtained from the leaves of Erythroxylon coca, a tree indigenous to Peru. For medical purposes, cocaine is applied topically or injected directly onto nerves to block pain transmission. Among its systemic actions is inhibition of the uptake of various brain neurotransmitters, including serotonin and dopamine. For this reason, cocaine is a central nervous system stimulant and euphoriant when taken orally, intravenously, or intranasally. For centuries, the leaves of this tree were chewed by native Peruvians to reduce fatigue and increase work capacity.[4] While the use of cocaine as a local anesthetic has been supplanted by other agents, it remains popular as a recreational drug.

Cannabis is the genus of plants that produces cannabinoids, the active chemicals in marijuana. The marijuana plant, which originated in Central Asia, was brought to the Americas by merchant traders.[5] Cannabis sativa is the most commonly cultivated species. The plant produces more than 60 different cannabinoids, the most mood-altering of which is delta-9-tetrahydrocannabinol. These chemicals induce their pharmacological effects by stimulating cannabinoid receptors, which are located on tissues throughout the body, including the brain. The endogenous activators of these receptors are referred to as endocannabinoids. Natural and synthetic cannabinoids have been examined extensively for their therapeutic potential.[6]

Cannabinoids are most commonly administered through the lungs by smoking the marijuana leaf. Marijuana is also consumed orally for recreational and medicinal purposes. A preparation of delta-9-tetrahydrocannabinol is available as a prescription drug. In addition to being euphoriants, cannabinoids are analgesics, antiemetics, and appetite stimulants. The cannabinoids are examples of natural products that were used historically for one effect, euphoria, but that are now being explored as potential drugs because of other pharmacological actions. It remains to be seen whether cannabinoids demonstrate any advantages over agents already available for these purposes.

Three of these four examples are alkaloids, the exception being the cannabinoids. Besides illustrating the rich variety of plant chemicals that influence brain function, this group of agents underscores the potential of alkaloids as agents for affecting brain function. Defining characteristics of alkaloids are that they are organic molecules that contain a nitrogen atom. It is notable that the majority of drugs used to treat central nervous system disorders are nitrogen-containing organic compounds as well. This structural similarity among psychoactive agents is no coincidence.

Plant alkaloids often make better drug candidates for treating central nervous system disorders than other types of plant constituents because of their pharmacokinetic and pharmacodynamic properties. With regard to the former, alkaloids tend to be more lipid soluble and resistant to drug metabolizing enzymes than, for example, flavonoids. These properties make the alkaloids more likely than many other plant products to cross from the intestinal tract into the blood, and from the blood into the brain. Their rate of metabolism and elimination is such that sufficient quantities of alkaloids can accumulate and remain in the brain for a period of time sufficient to induce noticeable central nervous system effects. As for pharmacodynamics, the alkaloid chemical structure is highly bioactive, especially with respect to brain neurotransmitter systems. This property may reflect the fact that plants manufacture alkaloids to repel insects and other predators. Thus, these plant products evolved because of their ability to influence the behavior of animals with nervous systems. While humans are not exempt from the toxic effects of alkaloids, they are fortunate that some of their therapeutically important central nervous system actions are achievable at doses below those that cause adverse effects. Given the large representation of alkaloids among the plant constituents known to affect brain function, the likelihood that a particular herbal product may modify central nervous system activity is increased if the active chemical is an alkaloid, or if the extract contains such compounds.

Knowledge of a relationship between a chemical class and the probability of an effect on the brain simplifies somewhat the assessment of herbal products purported to affect central nervous system activity. If the plant constituents are primarily flavonoids or other nonalkaloid compounds, the odds lengthen that the product includes agents with the pharmacokinetic and pharmacodynamic properties needed to influence brain function. This is especially true if the herbal product is touted as being free of side effects, as the vast majority of compounds that enter the brain have adverse effects at some dose. The lack of side effects is usually indicative of a lack of any effect. The presence of alkaloids can therefore be a criterion when assessing the validity of claims that an herbal supplement affects central nervous system activity.

In the end, there is no substitute for solid data, regardless of the chemical structure or chemical class of the compound thought to be responsible for the pharmacological effect. The consumer can be confident of a response if a pharmacologically active herbal constituent has been isolated and shown, preferably by several different investigators, to have the pharmacokinetic and pharmacodynamic properties required of a drug that influences brain function. While the absence of such data and even negative results do not prove the herb is devoid of central nervous system activity, the decision to use such a product would be based primarily on hope, not science. In that case, *caveat emptor* must be the guiding principle for the intelligent consumer, just as it should have been for Adam.

# Endnotes

## Chapter 1

[1] J. Lietava, "Medicinal plants in a Middle Paleolithic grave, Shanidar IV?" *Journal of Ethnopharmacology* 35 (1992): 263-266.

[2] R. C. Thompson, trans., *The Assyrian Herbal* (London: Luzac & Co., 1924).

[3] Pliny, *Natural History*, Bk.XXV, 42. trans. W.H.S. Jones. 2nd ed., (Cambridge, MA: Harvard University Press, 1956); Homer, *The Iliad* Bk. 11,840-848, trans. R. Fitzgerald, (Garden City, N.Y.,: Anchor Press, 1975).

[4] A. Heidel, *The Gilgamesh Epic and Old Testament Parallels*,(Chicago: University of Chicago Press, 1963).

[5] B.T. Moran, *Distilling Knowledge*, (Cambridge, MA: Harvard University Press, 2005).

[6] R. Boyle, *The Sceptical Chymist*, Reprint of the 1661 text (Mineola, NY: Dover Publications, Inc., 2003).

## Chapter 2

[1] J. Lietava, "Medicinal plants in a Middle Paleolithic grave, Shanidar IV?" *Journal of Ethnopharmacology* 35 (1992): 263-266.

[2] D. Csupor, Z. Petak, and J. Hohmann, "Medicinal properties of Hungarian Centauria species in the light of scientific evidence," *Acta Pharmaceutica Hungarica* 81 (2011): 63-75.

[3] A. Konakchiev, M. Todorova, B. Mikhova, A. Vitkova, and H. Najdenski, "Composition and antimicrobial activity of Achillea distans essential oil," *Natural Products Communications* 6 (2011): 905-906.

[4] H.-M. Chang and P. P.-H. But, eds., *Pharmacology and Applications of Chinese Materia Medica* trans. S.-C. Yao, L.-L. Wang, and S. S.-C. Yeung, (Singapore: World Scientific Publishing Co., 1986).

[5] P. Antolin, A. Girotti, F. J. Arias, B. Barriuso, P. Jimenez, M. A. Rojo, and T. Girbes, "Bacterial expression of biologically active recombinant musarmin 1 from bulbs of Muscari armeniacum L. and Miller," *Journal of Biotechnology* 112 (2004): 313-322.

[6] A. Deters. J. Zippel, N. Hellenbrand, D. Pappai, C. Possemeyer, and A. Hensel, "Aqueous extracts and polysaccharides from marshmallow roots (Althea officinalis L.): cellular internalisation and stimulation of cell physiology of human epithelial cells in vitro," *Journal of Ethnopharmacology* 127 (2010): 62-69.

[7] M. R. Lee, "The history of Ephedra (ma-huang)," *Journal of the Royal College of Physicians of Edinburgh* 41 (2011): 78-84.

[8] C. Wiart, *Ethnopharmacology of Medicinal Plants,* (Totowa, NJ: Humana Press, Inc., 2006).

[9] B. Ebell, trans., *The Papyrus Ebers,* (Copenhagen: Levin and Munksgaard, 1937).

[10] Theophrastus, *Concerning Odours, Enquiry into Plants, Vol. II,* A. F. Hort, trans. (Cambridge, MA: Harvard University Press, 1980).

[11] P. Dolara, B. Corte, C. Ghelardini, A. M. Pugliese, E. Cerbai, S. Menichetti, and A. Lo Nostro, "Local anesthetic, antibacterial and antifungal properties of sesquiterpenes from myrrh," *Planta Medica* 66 (2000): 356-358.

[12] T. M. Davis, *The Tomb of Hatshopsitu* (London: Gerald Duckworth and Co., 2004).

[13] R. C. Thompson, *The Assyrian Herbal: a Monograph*, Facsimile from UMI Books on Demand (Ann Arbor, MI: ProQuest Co., 2005).

[14] L. Clendening, *Source Book of Medical History* (New York: Dover Publications, Inc., 1960).

[15] J. Gerard, *The Herbal or General History of Plants* 1633 edition revised by Thomas Johnson (New York: Dover Publications, Inc., 1975).

[16] C. Celsus, *De Medicina*, W. G. Spencer, trans. (Cambridge, MA: Harvard University Press, 1971).

[17] S. Norton, "The Pharmacology of Mithridatum: a 2000-year-old Remedy," *Molecular Interventions* 6 (2006): 60-66.

[18] *Medicina Antiqua. Codex Vindobonensis 93, Vienna*, P. M. Jones, ed. (London: Harvey Miller Publishers, 1999).

[19] C. J. S. Thompson, *Alchemy and Alchemists* (Mineola, NY: Dover Publications, Inc., 2002).

[20] R. H. Major, *Classic Descriptions of Disease*, 3rd edition (Springfield, IL: Charles C. Thomas, Publisher, 1945).

[21] I. Gagliardi, *Li Trofei della Croce. L'Esperienza Gesuata; La Societa Lucchese tra Medievo de Eta Moderna* (Rome: Edizione di Storia e Letteratura, 2005).

[22] G. Andrea, *Libro de i Secretti con Ricette*, unpublished manuscript dated 1562, S. Norton, trans. (Lawrence, KS: The Kenneth Spencer Research Library, University of Kansas, trans. 2009).

[23] Paracelsus (Theophrastus von Hohenheim), *Four Treatises*, H. E. Sigerist, ed. (Baltimore: The Johns Hopkins University Press, 2005).

[24] B. T. Moran, *Distilling Knowledge: Alchemy, Chemistry and the Scientific Revolution* (Cambridge, MA: Harvard University Press, 2005).

[25] A. G. Debus, *The Chemical Philosophy* (Mineola, NY: Dover Publications, Inc., 2002).

[26] R. Boyle, *The Sceptical Chymist,* the classical 1661 text (Mineola, NY: Dover Publications, Inc., 2003).

[27] P. Strathern, *Mendeleyev's Dream: the Quest for the Elements,* (New York: St. Martin's Press, 2001).


## Chapter 3


[1] H. B. Murphree, "Narcotic analgesics, I: opium alkaloids," *Drill's Pharmacology in Medicine*, 3rd edition, J. R. DiPalma, editor, (New York: McGraw Hill Book Co., 1965).

[2] M. Gates and G. Tschudi, "Synthesis of morphine," *Journal of the American Chemical Society,* 78 (1956): 1380-1393.

[3] G. Andrea, *Libro de I Secretti con Ricette,* S. Norton, trans., unpublished manuscript, (Lawrence, KS: The Kenneth Spencer Research Library, University of Kansas, trans. 2009)

[4] C. Davison and H. G. Mandel, "Non-narcotic analgesics and antipyretics I: Salicylates," *Drill's Pharmacology in Medicine*, 3rd edition, J. R. DiPalma, ed. (New York: McGraw Hill Book Co., 1965).

[5] F. H. Kasten, "Paul Ehrlich: pathfinder in cell biology, I. Chronicle of his life and accomplishments in immunology, cancer research and chemotherapy," *Biotechnic and Histochemistry* 71 (1996): 2-37.

[6] A. H. Maehle, "Receptive substances: John Newport Langley (1852-1925) and his path to a receptor theory of drug action," *Medical History* 48 (2004): 153-174.

[7] N. Wiener, "Atropine, scopolamine and related antimuscarinic drugs," *Goodman and Gilman's, The Pharmacological Basis of Therapeutics*, 6th edition, A. G. Gilman, L. S. Goodman and A. Gilman, eds. (New York: Macmillan Publishing Co., Inc., 1980).

[8] A. J. Clark, *The Mode of Action of Drugs on Cells* (Baltimore: Williams and Wilkins, Co., 1933).

[9] H. I. Yamamura, S. J. Enna, and M. J. Kuhar, eds., *Neurotransmitter Receptor Binding*, 2nd edition (New York: Raven Press, 1985), 242 pp.

[10] N. Nash, "Overview of receptor cloning," *Current Protocols in Pharmacology,* S. J. Enna and M. Williams, eds. (New York, John Wiley & Sons, 1998), Unit. 6.1.

## Chapter 4

[1] J. W. Papez, "A proposed mechanism for emotion," *Archives of Neurology and Psychiatry* 38 (1937): 725-743.

[2] W. Zeman and J.R.M. Innes, eds., *Craigie's Neuroanatomy of the Rat,* (New York: Academic Press, 1963.

[3] S. Ferre, "An update on the mechanisms of the psychostimulant effects of caffeine," *Journal of Neurochemistry* 105 (2008): 1067-1079.

[4] A Saulin, M. Savli, and R. Lanzenberger, "Serotonin and molecular neuroimaging in humans using PET," *Amino Acids* 2011 e. pub 21947614.

[5] H. W. H. Tsang and T. Y. C. Ho, "A systematic review on the anxiolytic effects of aroma therapy on rodents under experimentally induced anxiety models," *Reviews in the Neurosciences* 21 (2010): 141-152.

[6] Y. Watanabe, A. Tsujimura, K. Takao, K. Nishi, Y. Ito, Y. Yasuhara, Y. Nakatomi, C. Yokoyama, K. Fukui, T. Miyakawa, and M. Tanaka, "Relaxin-3-deficient mice showed slight alteration in anxiety-related behavior." *Frontiers in Behavioral Neuroscience* 5 (2011): 50-65.

[7] K. Tamada, S. Tomonaga, F. Hatanaka, N. Nakai, K. Tatao, T. Miyakawa, J. Nakatani, and T. Takumi, "Decreased exploratory activity in a mouse model of 15q duplication syndrome: implications for the disturbance of serotonin signaling," PLoS ONW5 (2010): E15126.

[8] M. S. Kashani, M. R. Tavirani, S. A. Talaei, and M. Salami, "Aqueous extract of lavender (Lavandula angustifolia) improves the spatial performance of a rat model of Alzheimer's disease," *Neuroscience Bulletin* 27 (2011): 99-106.

[9] R. D'Hooge and P. P. De Deyn, "Applications of the Morris water maze in the study of learning and memory," *Research Reviews* 36 (2001): 60-90.

[10] D. H. Kim, H. A. Jung, S. J. Park, J. M. Kim, S. Lee, J. S. Choi, J. H. Cheong, K. H. Ko, and J. H. Ryu, "The effects of daidzin and its aglycon, daidzein, on the scopolamine-induced memory impairment in male mice," *Archives of Pharmacal Research* 33 (2010): 1685-1690.

[11] T. Karl, R. Pabst, and S. von Horsten, "Behavioral phenotyping of micc in pharmacological and toxicological research," *Experimental Toxicology and Pathology* 55 (2003): 69-83.

[12] A. Ramos, "Animal models of anxiety: do I need multiple tests?" *Trends in Pharmacological Science* 29 (2008): 493-498.

[13] J. F. Cryan and F. F. Sweeney, "The age of anxiety: role of animal models of anxiolytic action in drug discovery," *British Journal of Pharmacology* 164 (2011): 1129-1161.

[14] E. B. Olivieri, C. Vecchiato, N. Ignaccolo, P. Mannu, A. Castagna, L. Aravagli, V. Fontani, and S. Rinaldi, "Radioelectric brain stimulation in the treatment of generalized anxiety disorder with comorbid major depression in a psychiatric hospital: a pilot study," *Neuropsychiatric Disease and Treatment* 7 (2011): 449-455.

[15] I. Romera, V. Perez, J. M. Menchon, P. Polavieja, and I. Gilaberte, "Optimal cut-off point of the Hamilton Rating Scale for Depression according to normal levels of social and occupational functioning," *Psychiatry Research* 186 (2011): 133-137.

[16] S. Lehrl, "Clinical efficacy of kava extract WS 1490 in sleep disturbances associated with anxiety disorders. Results of a multicenter, randomized, placebo-controlled, double-blind clinical trial," *Journal of Affective Disorders* 78 (2004): 101-110.

## Chapter 5

[1] P. Henderson, *Handbook of Plants*, 2nd edition (New York: Peter Henderson & Co., 1904): 390.

[2] Li Shih-Chen, *Chinese Medicinal Herbs: A Modern Edition of a Classic Sixteenth Century Manual*, F. Porter Smith and G. A. Stuart, trans. (New York: Dover Publications, Inc., 2003).

[3] H.-M. Chang and P. P.-H. But, *Pharmacology and Applications of Chinese Materia Medica*, trans. S.-H. Yao, L.-L. Wang, and S. O.-S. Yeung, Vol. II (New Jersey: World Scientific Publishing Co., 1987): 1155.

[4] B. I. Koerner, "Ginkgo biloba? Forget about it," *Slate*, April 25, 2007.

[5] T. A. van Beek and P. Montoro, "Chemical analysis and quality control of *Ginkgo biloba* leaves, extracts, and phytopharmaceuticals," *Journal of Chromatography A* 1216 (2009): 2002-2032.

[6] A. Gawron-Gzella, P. Marek, J. Chanaj, and I. Matlawska, "Comparative analysis of pharmaceuticals and dietary supplements containing extracts from the leaves of Ginkgo biloba L," *Acta Poloniae Pharmaceutica* 67 (2010): 335-343.

[7] F. V. DeFeudis and K. Drieu, "Ginkgo biloba extract (EGb 761) and CNS functions: Basic studies and clinical applications," *Current Drug Targets* 1 (2000): 25-58.

[8] P. G. Pietta, C. Gardana, P. L. Mauri, R. Maffei-Facino, and M. Carini, "Identification of flavonoid metabolites after oral administration to rats of a Ginkgo biloba extract." *Journal of Chromatography B: Biomedical Applications* 673 (1995): 75-80.

[9] L. Rangel-Ordonez, M. Noldner, M. Schubert-Zsilavecz, and M. Wurglics, "Plasma levels and distribution of flavonoids in rat brain after single and repeated doses of standardized Ginkgo biloba extract EGb 761®," *Planta Medica* 76 (2010): 1683-1690.

[10] D. Tang, X. Yin, Z. Zhang, Y. Gao, Y. Wei, Y. Chen, and L. Han, "Comparative study on the pharmacokinetics of *Ginkgo biloba* extract between normal and diabetic rats by HPLC-DAD," *Latin American Journal of Pharmacy* 28 (2009): 400-408.

[11] P. L. Mauri, P. Simonetti, C. Gardana, M. Minoggio, P. Morazzoni, E. Bombardelli, and P. Pietta, "Liquid chromatography / atmospheric pressure chemical ionization mass spectrometry of terpene lactones in plasma of volunteers dosed with Ginkgo biloba L. extracts." *Rapid Communications in Mass Spectrometry* 15 (2001): 929-934.

[12] C. Ude, A. Paulke, M. Noldner, M. Schubert-Zsilavecz, and M. Wurglics, "Plasma and brain levels of terpene trilactones in rats after an oral single dose of standardized Ginkgo biloba extract EGb 761®," *Planta Medica* 77 (2011); 259-264.

[13] B. H. Hellum, Z. Hu, and O. G. Nilsen, "Trade herbal products and induction of CYP2C19 and CYP2E1 in cultured human hepatocytes," *Basic & Clinical Pharmacology & Toxicology* 105 (2009), 58-63.

[14] A. J. Lau and T. K. H. Chang, "Inhibition of human CYP2B6-catalyzed bupropion hydroxylation by Ginkgo biloba extract: Effect of terpene trilactones and flavonols," *Drug Metabolism and Distribution* 37 (2009); 1931-1937.

[15] O. Q. Yin, B. Tomlison, M. M. Wayne, A. H. Chow, and M. S. Chow, "Pharmacogenetics and herb-drug interactions: Experience with Ginkgo biloba and omeprazole," *Pharmacogenetics* 14 (2004), 841-850.

[16] S. Bastianetto, C. Ramassamy, S. Dore, Y. Christen, J. Poirier, and R. Quirion, "The Ginkgo biloba extract (EGB 761) protects hippocampal neurons against cell death induced by beta-amyloid," *European Journal of Neuroscience* 12 (2000), 1882-1890.

[17] M. Zielinska, A. Kostrzewa, E. Ignatowicz, and J. Budzianowski, "The flavonoids, quercetin and isorhamnetin 3-O-acylglucosides diminish neurophil oxidative metabolism and lipid peroxidation," *Acta Biochimica Polonica* 48 (2001); 183-189.

[18] K. F. Chung, M. McCusker, C. P. Page, G. Dent, Ph. Guinot, and P. J. Barnes, "Effect of ginkgolide mixture (BN 52063) in antagonizing skin and platelet responses to platelet activating factor in man," *The Lancet* January 31 (1987), 248-250.

[19] D. Nunez, M. Chignard, R. Korth, J. P. Couedic, X. Norel, B. Spinnewyn, P. Braquet, and J. Benveniste, "Specific inhibition of PAF-acether-induced platelet activation by BN 52021 and comparison with the PAF-acether inhibitors kadsurenone and CV 3988," *European Journal of Pharmacology* 123 (1986), 197-205.

[20] F. Jung, C. Mrowietz, H. Kiesewetter, and E. Wenzel, "Effect of Ginkgo biloba on fluidity of blood and peripheral microcirculation in volunteers," *Arzneimittelforschung* 40 (1990), 589-593.

[21] H. H. Dodge, T. Zitzelberger, B. S. Oken, D. Howierson, and J. Kaye, "A randomized placebo-controlled trial of Ginkgo biloba for prevention of cognitive decline," *Neurology* 70 (2008): 19, Pt.2: 1809-1817.

22 B. E. Snitz, E. S. O'Meara, M. C. Carlson, A. M. Arnold, D. G. Ives, S. R. Rapp, J. Saxton, O. L. Lopez, L. O. Dunn, K. M. Simb, and S. T. DeKosky, "Ginkgo biloba for preventing cognitive decline in older adults: a randomized trial," *Journal of the American Medical Association* 302 (2009): 24: 2663-2670.

23 S. T. DeKosky, J. D. Williamson, A. L. Fitzpatrick, R. A. Kronmal, D. G. Ives, J. A. Saxton, O. L. Lopez, G. Burke, M. C. Carlson, L. P. Fried, L. H. Kuller, J. A. Robbins, R. P. Tracy, N. F. Woolard, L. Dunn, B. E. Snitz, R. L. Nahin, C. D. Furberg, and Ginkgo Evaluation of Memory (GEM) Study Investigators, "Ginkgo biloba for prevention of dementia: a randomized controlled trial," *Journal of the American Medical Association* 300 (2008), 2253-2262.

24 A. Dorak, O. Uzun, and P. Ozsohin, "A placebo-controlled study of extract of Ginkgo biloba added to clozapine in patients with treatment-resistant schizophrenia," *International Clinical Psychopharmacology* 23 (2008): 4: 223-227.

25 G. D. Gardner, J. L. Zehnder, A. J. Rigby, J. R. Nicholaus, and J. W. Farquhar, "Effect of Ginkgo biloba (EGb 761) and aspirin on platelet aggregation and platelet function analysis among older adults at risk of cardiovascular disease: a randomized clinical trial," *Blood Coagulation and Fibrinolysis* 18 (2007):8: 787-793.

## Chapter 6

1 J. Gerard, *The Herbal or General History of Plants*, 1633 edition (New York: Dover Publications, Inc., 1975).

2 Pliny, *Natural History*, Book XXVI, Section 158, trans. W. H. S. Jones, 2nd edition (Cambridge, MA: Harvard University Press, 1980).

3 Li Shih-Chen, *Chinese Medicinal Herbs: A Modern Edition of a Classic Sixteenth-Century Manual,* trans. F. Porter Smith and G. A. Stuart (New York: Dover Publications, Inc., 2003).

4 *Medicina Antiqua*, Codex Vindobonensis 93, Vienna. Folio 151V. Manuscript facsimile (London: Harvey Miller Publishers, 1999).

5 G. Andrea, *Libro de i Secreti con Ricette*, unpublished manuscript dated 1562, S. Norton, trans. (Lawrence, KS: The Kenneth Spencer Research Library, University of Kansas, trans. 2009).

6 M. Grieve, *A Modern Herbal* (New York: Barnes & Noble, Inc., 1996).

7 E. Ernst, "St. John's wort: an anti-depressant? A systematic, criteria-based review." *Phytomedicine* 2 (1995): 67-71.

8 Z. Sadigge, I. Nalem, and A. Maimoona, "A review of the antibacterial activity of Hypericum perforatum L." *Journal of Ethnopharmacology* 131 (2010): 511-521.

9 R. Kerb, J. Brockmaller, B. Staffeldt, M. Ploch, and I. Roots, "Single-dose and steady state pharmacokinetics of hypericin and pseudohypericin," *Antimicrobial Agents and Chemotherapy* 40 (1996): 2087-2193.

[10] A. Biber, H. Fisher, A. Romer, and S. S. Chatterjee, "Oral bioavailability of hyperforin from Hypericum extracts in rats and human volunteers," *Pharmacopsychology* 31 (1998): 36-43.

[11] G. Di Carlo, F. Borrelli, E. Ernst, and A. A. Izzo, "St John's wort: Prozac from the plant kingdom," *Trends in Pharmacological Sciences* 22 (2001): 292-297.

[12] S. Caccia and M. Gobbi, "St. John's wort components and the brain: uptake, concentrations reached and the mechanisms underlying pharmacological effects," *Current Drug Metabolism* 10 (2009): 1055-1065.

[13] M. Wurglics, M. Schubert-Zsilavecz, "Hypericum perforatum: a 'modern' herbal antidepressant: pharmacokinetics of active ingredients," *Clinical Pharmacokinetics* 45 (2006): 449-468.

[14] S. Caccia, "Antidepressant-like components of Hypericum perforatum extracts: an overview of their pharmacokinetics and metabolism," *Current Drug Metabolism* 6 (2005): 531-543.

[15] J. Hokkanen, A. Tolonen, S. Mattila, and M. Turpeinen, "Metabolism of hyperforin, the active constituent of St. John's wort, in human liver microsomes," *European Journal of Pharmaceutical Sciences* 42 (2011): 273-284.

[16] V. Di Matteo, G. Di Giovanni, M. Di Mascio, and E. Esposito, "Effect of acute administration of Hypericum perforatum-$CO_2$ extract on dopamine and serotonin release in the rat central nervous system," *Pharmacopsychology* 33 (2000): 14-18.

[17] J. Barnes, L. A. Anderson, and J. D. Phillipson, "St. John's wort (Hypericum perforatum L.): a review of its chemistry, pharmacology and clinical properties," *Journal of Pharmacy and Pharmacology* 53 (2001): 583-590.

[18] A. Singer, M. Wonneman, and W. E. Muller, "Hyperforin, a major antidepressant constituent of St. John's wort, inhibits serotonin uptake by elevating free intracellular $Na^+$," *Journal of Pharmacology and Experimental Therapeutics* 290 (1999): 1363-1368.

[19] K. Leuner, V. Kazanski, M. Muller, K. Essin, B. Henke, M. Gollasch, C. Harteneck, and W. E. Muller, "Hyperforin—a key constituent of St. John's wort specifically activates TRPC6 channels," *Federation of American Societies for Experimental Biology Journal* 21 (2007): 4101-4111.

[20] L. A. Chahl, "TRP Channels and Psychiatric Disorders," *Advances in Experimental Medicine and Biology* 704 (2011): 987-1009.

[21] T. Hayashi, S. Y. Tsai, T. Mori, M. Fujimoto, and T. P. Su, "Targeting ligand-operated chaperone sigma-1 receptors in the treatment of neuropsychiatric disorders," *Expert Opinion on Therapeutic Targets* 15 (2011): 557-577.

[22] T. Mennini and M. Gobbi, "The antidepressant mechanism of Hypericum perforatum," *Life Sciences* 75 (2004): 1021-1027.

[23] B. Kraus, H. Wolff, E. F. Elstner, and J. Heilmann, "Hyperforin is a modulator of inducible nitric oxide synthase and phagocytosis in microglia and macrophages," *Naunyn Schmieberger Archive Pharmacologia* 381 (2010): 541-553.

24 R. Crupi, E. Mazzon, A. Marino, G. La Spada, P. Bramanti, F. Battaglia, S. Cuzzocrea, and E. Spina, "Hypericum perforatum treatment: effect on behaviour and neurogenesis in a chronic stress model in mice," *BMC Complementary Alternative Medicine* 11 (2011): 7.

25 S. S. Chatterjee, S. K. Bhattacharya, M. Wonnemann, A. Singer, and W. E. Muller, "Hyperforin as a possible antidepressant component of hypericum extracts," *Life Sciences* 63 (1998): 499-510.

26 K. Linde, G. Ramirez, C. D. Mulrow, A. Paulo, W. Weidenhammer, and D. Melchart, "St. John's wort for depression—an overview and meta-analysis of randomized clinical trials," *British Medical Journal* 313 (1996): 253-258.

27 H. L. Kim, J. Streltzer, and D. Goebert, "St. John's wort for depression: a meta-analysis of well defined clinical trials," *Journal of Nervous and Mental Disease* 187 (1999): 532-539.

28 S. Kasper, F. Caraci, B. Forti, F. Drago, and E. Aguglia, "Efficacy and tolerability of Hypericum extract for the treatment of mild to moderate depression," *European Neuropsychopharmacology* 20 (2010): 747-765.

29 K. Linde, M. M. Berner, and L. Kriston, "St. John's wort for major depression," *Cochrane Database of Systematic Reviews* (2008), Issue 4. Art. No.: CD000448. DOI:10.1002/14651858.CD000448.pub3.

30 Y. Lecrubier, G. Clerc, R. Didi, and M. Kieser, "Efficacy of St. John's wort extract WS 5570 in major depression: a double-blind, placebo-controlled trial," *American Journal of Psychiatry* 159 (2002): 1361-1366.

31 H. Woelk, "Comparison of St. John's wort and imipramine for treating depression: randomized controlled trial," *British Medical Journal* 321 (2000): 536-539.

32 M. Fava, J. Alpert, A. A. Nierenberg, D. Mischoulon, M. W. Otto, J. Zajecka, H. Murck, and J. F. Rosenbaum, "A double-blind, randomized trial of St. John's wort, fluoxetine, and placebo in major depressive disorder," *Journal of Clinical Psychopharmacology* 25 (2005): 441-447.

33 R. C. Shelton, M. B. Keller, A. Gelenberg, D. L. Dunner, R. Hirschfeld, M. E. Thase, J. Russell, R. B. Lydiard, P. Crits-Cristoph, R. Gallop, L. Todd, D. Hellerstein, P. Goodnick, G. Keitner, S. M. Stahl, and U. Halbreich, "Effectiveness of St. John's wort in major depression: a randomized controlled trial," *Journal of the American Medical Association* 285 (2001): 1978-1986.

34 Hypericum Depression Study Group, "Effect of hypericum perforatum (St. John's wort) in major depressive disorder: a randomized controlled trial," *Journal of the American Medical Association* 287 (2002): 1807-1814.

35 K. D. Trautmann-Sponsel and A. Dienel, "Safety of Hypericum extract in mildly to moderately depressed outpatients: a review based on data from three randomized, placebo-controlled trials," *Journal of Affective Disorders* 82 (2004): 303-307.

36 D. Durr, B. Stieger, G. A. Kullak-Ublick, K. M, Rentsch, H. C. Steinert, P. J. Meier, and K. Fattinger, "St. John's wort induces intestinal P-glycoprotein/MDR1 and intestinal and hepatic CYP34A," *Clinical Pharmacology and Therapeutics* 68 (2000): 598-604.

37 F. Borrelli and A. A. Izzo, "Herb-drug interactions with St. John's wort (Hypericum perforatum): an update on clinical observations," *AAPS Journal* 11 (2009): 710-727.

38 R. Madabushi, B. Frank, B. Drewelow, H. Derendorf, and V. Butterweck, "Hyperforin in St. John's wort drug interactions," *European Journal of Clinical Pharmacology* 62 (2006): 225-233.

39 J. J. Dugouga, E. Mills, D. Perri, and G. Coren, "Safety and efficacy of St. John's wort (Hypericum) during pregnancy and lactation," *Canadian Journal of Clinical Pharmacology* 13 (2006): 268-276.

40 A. O. da Conceicao, L. Takser, and J. Lafond, "Effect of St. John's Wort standardized extract and hypericin on in vitro placental calcium transport," *Journal of Medicinal Food* 13 (2010): 934-942.

## Chapter 7

1 Li Shih-Chen, *Chinese Medicinal Herbs*, The Herbal Pen Ts'ao of 1578 translated by F. P. Smith and G. A. Stuart (New York: Dover Publications, Inc., 2003).

2 N. Culpeper, *Complete Herbal and English Physician*, reproduced from 1826 edition (Leicester, UK: Magna Books, 1992).

3 N. J. Jacobo-Herrera, N. Vartianinen, P. Bremner, S. Gibbons, J. Koistinaho, and M. Heinrich, "NF-kappa B modulators from Valeriana officinalis," *Phytotherapy Research* 20 (2006): 917-919.

4 T. D. Gilmore and M. Herscovitch, "Inhibitors of NF-kappa B signaling: 785 and counting," *Oncogene* 25 (2006): 6887-6899.

5 G. Cravotto, L. Boffa, L. Genzini, and D. Garella, "Phytotherapeutics: an evaluation of the potential of 1000 plants," *Journal of Clinical Pharmacology and Therapeutics* 35 (2010): 11-48.

6 J. Gerard, *The Herbal or General History of Plants*, reprint of 1633 edition. (New York: Dover Publications, Inc., 1975).

7 Theophrastus, *Enquiry into Plants*, trans., A. Hort (Cambridge, MA: Harvard University Press, 1999).

8 X. Q. Gao and L. Bjork, "Valerenic acid derivatives and valepotriates among individuals, varieties and species of Valeriana," *Fitoterapia* 71 (2000): 19-24.

9 D. Shohet, R. B. Wills, and D. L. Stuart, "Valepotriates and valerenic acid in commercial preparations of valerian available in Australia," *Pharmazie* 56 (2001): 860-863.

10 J. G. Ortiz, N. Rassi, P. M. Maldonado, S. Gonzalez-Cabrera, and I. Ramos, "Commercial valerian interactions with [3H]flunitrazepam and [3H]MK-801 binding in rat synaptic membranes," *Phytotherapy Research* 20 (2006): 794-798.

[11] P. J. Houghton, "The scientific basis for the reputed activity of Valerian," *Journal of Pharmacy and Pharmacology* 51 (1999): 505-512.

[12] G. D. Anderson, G. W. Elmer, E. D. Kantor, I. E. Templeton, and M. V. Vitiello, "Pharmacokinetics of valerenic acid after administration of valerian in healthy subjects," *Phytotherapy Research* 19 (2005): 801-803.

[13] G. D. Anderson, G. W. Elmer, D. M. Taibi, M. V. Vitiello, E. Kantor, T. F. Kalhorn, W. N. Howald, S. Barsness, and C. A. Landis, "Pharmacokinetics of valerenic acid after single and multiple doses of valerian in older women," *Phytotherapy Research* 24 (2010): 1442-1446.

[14] W. Neuhaus, G. Trauner, D. Gruber, S. Oelzant, W. Klepal, B. Kopp, and C. R. Noe, "Transport of a $GABA_A$ receptor modulator and its derivatives from Valeriana officinalis L.s.l. across an in vitro cell culture model of the blood-brain barrier," *Planta Medica* 74 (2008): 1338-1344.

[15] U. Simmen, C. Saladin, P. Kaufmann, M. Poddar, C. Wallimann, and W. Schaffner, "Preserved pharmacological activity of hepatocytes-treated extracts of valerian and St. John's wort," *Planta Medica* 71 (2005): 592-598.

[16] A. Maier-Salamon, G. Trauner, R. Hiltscher, G. Reznicek, B. Kopp, T. Thalhammer, and W. Jager, "Heptatic metabolism and biliary excretion of valerenic acid in isolated perfused rat livers: role of Mrp2 (Abcc2)," *Journal of Pharmaceutical Science* 98 (2009): 3839-3849.

[17] B. H. Hellum and O. G. Nilsen, "In vitro inhibition of CYP3A4 metabolism and P-glycoprotein-mediated transport by trade herbal products," *Basic Clinical Pharmacology and Toxicology* 102 (2008): 466-475.

[18] J. L. Donovan, C. L. DeVane, K. D. Chavin, J. S. Wang, B. B. Gibson, H. A. Gefroh, and J. S. Markowitz, "Multiple night-time doses of valerian (Valeriana officinalis) had minimal effects on CYP3A4 activity and no effect on CYP2D6 activity in healthy volunteers," *Drug Metabolism and Disposition* 32 (2004): 1333-1336.

[19] H. Mohler, "Functional relevance of $GABA_A$-receptor subtypes," in *The Receptors: The GABA Receptors, Third Edition*, S. J. Enna and H. Mohler, eds. (Totowa, NJ: Humana Press, Inc., 2007), pp. 23-39.

[20] C. Wasowski, M. Marder, H. Viola, J. H. Medina, and A. C. Paladini, "Isolation and identification of 6-methylapigenin, a competitive ligand for the brain GABA(A) receptors, from Valeriana wallichii," *Planta Medica* 68 (2002): 934-936.

[21] R. E. Granger, E. L. Campbell, and G. A. Johnston, "(+)- and (-)- borneol: efficacious positive modulators of GABA action at human recombinant alpha1beta2gamma2L GABA(A) receptors," *Biochemical Pharmacology* 69 (2005): 1101-1111.

[22] B. Schumacher, S. Scholle, J. Holzi, N. Khudeir, S. Hess, and C. E. Muller, "Lignans isolated from valerian: identification and characterization of a new olivil derivative with partial agonistic activity at A(1) adenosine receptors," *Journal of Natural Products* 65 (2002): 1479-1485.

[23] B. M. Dietz, G. B. Mahady, G. F. Pauli, and N. R. Farnsworth, "Valerian extract and valerenic acid are partial agonists of the 5-HT5a receptor in vitro," *Brain Research Molecular Brain Research.* 138 (2005): 191-197.

[24] S. Khom, I. Baburin, E. Timin, A. Hohaus, G. Trauner, B. Kopp, and S. Hering, "Valerenic acid potentiates and inhibits GABA(A) receptors: molecular mechanism and subunit specificity," *Neuropharmacology* 53 (2007): 178-187.

[25] G. Trauner, S. Kohm, I. Baburin, B. Benedek, S. Hering, and B. Kopp, "Modulation of GABAA receptor by valerian extracts is related to the content of valerenic acid," *Planta Medica* 74 (2008): 19-24.

[26] D. Benke, A. Barberis, S. Kopp, K. H. Altmann, M. Schubiger, K. E. Vogt, U. Rudolph, and H. Mohler, "GABA A receptors as in vivo substrate for the anxiolytic action of valerenic acid, a major constituent of valerian root extracts," *Neuropharmacology* 56 (2009): 174-181.

[27] S. Yuan, "The gamma-aminobutyric acidergic effects of valerian and valerenic acid on rat brainstem neuronal activity," *Anesthesia and Analgesia* 98 (2004): 353-358.

[28] H. Hendriks, R. Bos, H. J. Woerdenbag, A. S. Koster, "Central nervous depressant activity of valerenic acid in the mouse," *Planta Medica* 51 (1985): 28-31.

[29] S. Fernandez, C. Wasowski, A. C. Paladini, and M. Marder, "Sedative and sleep-enhancing properties of linarin, a flavonoid isolated from Valeriana officinalis," *Pharmacology, Biochemistry and Behavior* 77 (2004): 399-404.

[30] K. Murphy, Z. J. Kubin, J. N. Shepherd, and R. H. Ettinger, "Valeriana officinalis root extracts have potent anxiolytic effects in laboratory rats," *Phytomedicine* 17 (2010): 674-678.

[31] S. Bent, A. Padula, D. Moore, M. Patterson, and W. Mehling, "Valerian for sleep: a systematic review and meta-analysis," *American Journal of Medicine* 119 (2006): 1005-1012.

[32] M. I. Fernandez-San-Martin, R. Masa-Font, L. Palacios-Soler, P. Sancho-Gomez, C. Calbo-Caldentey, and G. Flores-Mateo, "Effectiveness of Valerian on insomnia: a meta-analysis of randomized placebo-controlled trials," *Sleep Medicine* 11 (2010): 505-511.

[33] D. L. Barton, P. J. Atherton, B. A. Bauer, D. F. Moore, B. I. Mattar, B. I. Lavasseur, K. M. Rowland, R. T. Zon, N. A. Lilindqwister, G. G. Nagargoje, T. I. Morgenthaler, J. A. Sloan, and C. L. Loprinzi, "The use of valeriana officinalis (valerian) in improving sleep in patients who are undergoing treatment for cancer: a phase III randomized, placebo-controlled, double-blind study: NCCTG Trial, N01C5," *Journal of Supportive Oncology* 9 (2011): 24-31.

[34] B. J. Isetts, "Valerian," Chapter 4 in *Herbal Products: Toxicology and Clinical Pharmacology*, T. S. Tracy and R. L. Kingston, eds., 2nd edition (Totowa, NJ: Humana Press, 2007).

35 J. L. Donovan, C. L. De Vane, K. D. Chavin, J. S. Wang, B. B. Gibson, H. A. Gefroh, and J. S. Markowitz, "Multiple night-time dose of valerian (Valerian officinalis) had minimal effects on CYP3A4 activity and no effect on CYP2D6 activity in healthy volunteers," *Drug Metabolism and Disposition* 32 (2004): 1333-1336.

## Chapter 8

1 J. Gerard, *The Herbal or General History of Plants*, reprint of 1633 edition (New York: Dover Publications, Inc., 1975).

2 D. Wheatley, "Medicinal Plants for insomnia: a review of their pharmacology, efficacy and tolerability," *Journal of Psychopharmacology* 19 (2005): 414-421.

3 E. Perry and M. Howes, "Medicinal plants and dementia therapy: herbal hopes for brain aging?" *CNS Neuroscience and Therapeutics* 17 (2011): 683-698.

4 J. Patora, T. Majda, J. Gora, and B. Klimek, "Variability in the content and composition of essential oil from lemon balm (Melissa officinalis L.) cultivated in Poland," *Acta Poloniae Pharmaceutica* 60 (2003): 395-400.

5 B. Uysal, F. Sozmen, and B. S. Buyuktas, "Solvent-free microwave extraction of essential oils from Laurus nobilis and Melissa officinalis: comparison with conventional hydro-distillation and ultrasound extraction," *Natural Products Communications* 5 (2010): 111-114.

6 M. Mrlianova, D. Tekelova, M. Felklova, V. Reinohl, and J. Toth, "The influence of the harvest cut height on the quality of the herbal drugs Melissae folium and Melissae herba," *Planta Medica* 68 (2002): 178-180.

7 R. Awad, A. Muhammad, T. Durst, V. L. Trudeau, and J. T. Arnason, "Bioassay-guided fractionation of lemon balm (Melissa officinalis L.) using an in vitro measure of GABA transaminase activity, *Phytotherapy Research* 23 (2009): 1075-1081.

8 A. Ibarra, N. Feuillere, M. Roller, E. Lesburgere, and D. Beracochea, "Effects of chronic administration of Melissa officinalis L. extract on anxiety-like reactivity and on circadian and exploratory activities in mice," *Phytomedicine* 17 (2010): 397-403.

9 J. Diliberto, G. Usha, and L. Birnbaum, "Disposition of citral in male Fischer rats," *Drug Metabolism and Disposition* 16 (1988): 721-727.

10 Q. Liao, W. Yang, Y. Jia, X. Chen, Q. Gao, and K. Bi, "LC-MS determination and pharmacokinetic studies of ursolic acid in rat plasma after administration of the traditional Chinese medicinal preparation Lu-Ying extract," *Yakugaku Zasshi* 125 (2005): 509-515.

11 Q. Chen, S. Luo, Y. Zhang, and Z. Chen, "Development of a liquid chromatography-mass spectrometry method for the determination of ursolic acid in rat plasma and tissue: Application to the pharmacokinetic and tissue distribution study," *Analytical and Bioanalytical Chemistry* 399 (2011): 2877-2884.

[12] Y. Liu, X. Li, Y. Li, L. Wang, and M. Xue, "Simultaneous determination of danshensu, rosmarinic acid, cryptotanshinone, tanshinone IIA, tanshinone I and dihydrotanshinone I by liquid chromatographic-mass spectrometry and the application to pharmacokinetics in rats," *Journal of Pharmaceutical and Biomedical Analysis* 53 (2010): 698-704.

[13] Y. Konishi, Y. Hitomi, M. Yoshida, and E. Yoshioka, "Pharmacokinetic study of caffeic and rosmarinic acids in rats after oral administration," *Journal of Agricultural and Food Chemistry* 53 (2005): 4740-4746.

[14] Y. Konishi and S. Kobayashi, "Transepithelial transport of rosmarinic acid in intestinal Caco-2 cell monolayers," *Bioscience Biotechnology and Biochemistry* 69 (2005): 583-591.

[15] S. Baba, N. Osakabe, M. Natsume, and J. Terao, "Orally administered rosmarinic acid is present as the conjugated and/or methylated forms in plasma, and is degraded and metabolized to conjugated forms of caffeic acid, ferulic acid and m-coumaric acid," *Life Sciences* 75 (2004): 165-178.

[16] T. Nakazawa and K. Ohsawa, "Metabolism of rosmarinic acid in rats," *Journal of Natural Products* 61 (1998): 993-996.

[17] H. Ji, B. Shin, D. Joeng, E. Park, S. Yoo, and H. Lee, "Interspecies scaling of oleanolic acid in mice, rats, rabbits and dogs and prediction of human pharmacokinetics," *Archives of Pharmacal Research* 32 (2009): 251-257.

[18] M. Song, T. Hang, Y. Wang, L. Jiang, X Wu, Z. Zhang, J. Shen, and Y. Zhang, "Determination of oleanolic acid in human plasma and study of its pharmacokinetics in Chinese healthy male volunteers by HPLC tandem mass spectrometry," *Journal of Pharmaceutical and Biomedical Analysis* 40 (2006): 190-196.

[19] N. Yoshida, A. Takagi, H. Kitazawa, J. Kawakami, and I. Adachi, "Inhibition of P-glycoprotein-mediated transport by extracts of and monoterpenoids contained in Zanthoxyli fructus," *Toxicology and Applied Pharmacology* 209 (2005): 167-173.

[20] N. Yoshida, A. Takagi, H. Kitazawa, J. Kawakami, and I. Adachi, "Effects of citronellal, a monoterpenoid in Zanthoxyli Fructus, on the intestinal absorption of digoxin in vitro and in vivo," *Journal of Pharmaceutical Sciences* 95 (2006): 552-560.

[21] J. Wen and Y. Xiong, "The effect of herbal medicine danshensu and ursolic acid on pharmacokinetics of rosuvastatin in rats," *European Journal of Drug Metabolism and Pharmacokinetics* 36 (2011): 205-211.

[22] K. Seo, H. Kim, H. Ku, H. Ahn, S. Park, S. Bae, J. Shin, and K. Liu, "The monoterpenoids citral and geraniol are moderate inhibitors of CYP2B6 hydroxylase activity," *Chemical and Biological Interactions* 174 (2008): 141-146.

[23] G. Guginski, A. P. Luiz, M. D. Silva, M. Massaro, D. F. Martins, J. Chaves, R. W. Mattos, D. Silveira, V. M. Ferreira, J. B. Calixto, and A. R. Santos, "Mechanisms involved in the antinociception caused by ethanolic extract obtained from the leaves of Melissa officinalis (lemon balm) in mice," *Pharmacology, Biochemistry and Behavior* 93 (2009): 10-16.

[24] G. Gamaro, E. Suyenaga, M. Borsoi, J. Lermen, P. Pereira, and P. Ardenghi, "Effect of rosmarinic and caffeic acids on inflammatory and nociception responses in rats," *ISRN Pharmacology* (2011): Epub 2011:451682, DOI: 10.5402/2011/451682.

[25] P. Pereira, D. Tysca, P. Oliveira, L. F. da Silva Brum, J. N. Picada, and P. Ardenghi, "Neurobehavioral and genotoxic aspects of rosmarinic acid," *Pharmacology Research* 52 (2005): 199-203.

[26] M. L. Ortiz, M. P. Gonzalez-Garcia, H. A. Ponce-Monter, G. Castaneda-Hernandez, and P. Aguilar-Robles, "Synergistic effect of the interaction between naproxen and citral on inflammation in rats," *Phytomedicine* 18 (2010): 74-79.

[27] M. S. Melo, L. C. Sena, F. J. Barreto, L. R. Bonjardim, J. R. Almeida, J. T. Lima, D. P. De Sousa, and L. J. Quintans-Junior, "Antinociceptive effect of citronellal in mice," *Pharmaceutical Biology* 48 (2010): 411-416.

[28] R. Awad, D. Levac, P. Cybulska, Z. Merali, V. L. Trudeau, and J. T. Arnason, "Effects of traditionally used anxiolytic botanicals on enzymes of the gamma-aminobutyric acid (GABA) system," *Canadian Journal of Physiology and Pharmacology* 85 (2007): 933-942.

[29] D. Y. Yoo, J. H. Choi, W. Kim, C. H. Lee, Y. S. Yoon, M. H. Won, and I. K. Hwang, "Effects of Melissa officinalis L. (lemon balm) extract on neurogenesis associated with serum corticosterone and GABA in the mouse dentate gyrus," *Neurochemical Research* 36 (2011): 250-257.

[30] K. Dastmalchi, V. Ollilainen, P. Lackman, G. Boije af Gennas, H. J. Dorman, P. P. Jarvinen, J. Yli-Kauhaluoma, and R. Hiltunen, "Acetylcholinesterase inhibitory guided fractionation of Melissa officinalis L," *Bioorganic and Medicinal Chemistry* 17 (2009): 867-871.

[31] D. O. Kennedy, G. Wake, S. Savelev, E. K. Perry, K. A. Wesnes, and A. B. Scholey, "Modulation of mood and cognitive performance following acute administration of single doses of Melissa officinalis (Lemon balm) with human CNS nicotinic and muscarinic receptor-binding properties," *Neuropsychopharmacology* 28 (2003): 1871-1881.

[32] T. Hamaguchi, K. Ono, A. Murase, and M. Yamada, "Phenolic compounds prevent Alzheimer's pathology through different effects on the amyloid-beta pathway," *American Journal of Pathology* 175 (2009): 2557-2565.

[33] V. Lopez, S. Martin, M. P. Gomez-Serranillos, M. E. Carretero, A. K. Jager, and M. L. Calvo, "Neuroprotective and neurological properties of Melissa officinalis," *Neurochemical Research* 34 (2009): 1955-1961.

[34] N. Mimica-Dukic, B. Bozin, M. Sokovic, and N. Simin, "Antimicrobial and antioxidant activities of Melissa officinalis L. (Lamiaceae) essential oil," *Journal of Agricultural Food Chemistry* 52 (2004): 2485-2489.

[35] M. J. Chung, S. Y. Cho, M. J. Bhuiyan, K. H. Kim, and S. J. Lee, "Anti-diabetic effects of lemon Balm (Melissa officinalis) essential oil on glucose- and lipid-regulating enzymes in type 2 diabetic mice," *British Journal of Nutrition* 104 (2010): 180-188.

[36] J. Liu, "Oleanoic acid and ursolic acid: research perspectives," *Journal of Ethnopharmacology* 22 (2005): 92-94.

[37] R. Gautam and S. M. Jachak, "Recent developments in anti-inflammatory natural products," *Medicinal Research Reviews* 29 (2009): 767-820.

[38] Y. Ikeda, A. Murakami, and H. Ohigashi, "Ursolic acid: an anti- and pro-inflammatory triterpenoid," *Molecular Nutrition and Food Research* 52 (2008): 26-42.

[39] M. H. Shyu, T. C. Kao, and G. C. Yen, "Oleanoic acid and ursolic acid induce apoptosis in HuH7 human hepatocellular carcinoma cells through a mitochondrial-dependent pathway and downregulation of XIAP," *Journal of Agricultural Food Chemistry* 58 (2010): 6110-6118.

[40] C. G. Ballard, J. T. O'Brien, K. Reichelt, and E. K. Perry, "Aromatherapy as a safe and effective treatment for the management of agitation in severe dementia: the results of a double-blind, placebo-controlled trial with Melissa," *Journal of Clinical Psychiatry* 63 (2002): 553-558.

[41] D. O. Kennedy and A. B. Scholey, "The psychopharmacology of European herbs with cognition-enhancing properties," *Current Pharmaceutical Design* 12 (2006): 4613-4623.

[42] A. Burns, E. Perry, C. Holmes, P. Francis, J. Morris, M. J. Howes, P. Chazot, G. Lees, and C. Ballard, "A double-blind placebo-controlled randomized trial of Melissa officinalis oil and donepezil for the treatment of agitation in Alzheimer's Disease," *Dementia and Geriatric Cognitive Disorders* 31 (2011): 158-164.

[43] D. O. Kennedy, W. Little, C. Haskell, and A. Scholey, "Anxiolytic effects of a combination of Melissa officinalis and Valeriana officinalis during laboratory induced stress," *Phytotherapy Research* 20 (2006): 96-102.

## Chapter 9

[1] C. Darwin, *The Voyage of the Beagle* (Vercelli, Italy: White Star Publishers, 2006).

[2] Y. N. Singh and M. Blumenthal, "Kava: an overview. Distribution, mythology, botany, culture, chemistry and pharmacology of the South Pacific's most revered herb," *HerbalGram* 39 (1997): 33-55.

[3] N. N. Singh, S. D. Singh, and Y. N. Singh, "Kava: Clinical studies and therapeutic implications," in Y. N. Singh, ed., *Kava from Ethnology to Pharmacology* (Boca Raton, FL: CRC Press, 2004).

[4] D. Wheatley, "Medicinal plants for insomnia: a review of their pharmacology, efficacy and tolerability," *Journal of Psychopharmacology* 19 (2005): 414-421.

[5] D. Wheatley, "Stress-induced insomnia treated with kava and valerian: singly and in combination," *Human Psychopharmacology* 16 (2001): 353-356.

[6] S. E. Lakhan and K. F. Vieira, "Nutritional and herbal supplements for anxiety and anxiety-related disorders: systematic review," *Nutrition Journal* 9 (2010): 42-58.

[7] I. Ramzan and V. H. Tran, "Chemistry of Kava and Kavalactones," in Y. N. Singh, ed., *Kava from Ethnology to Pharmacology* (Boca Raton, FL: CRC Press, 2004).

[8] V. Lebot and J. Levesque, "Genetic control of kavalactone chemotypes in Piper methysticum cultivars," *Phytochemistry* 43 (1996): 397-405.

[9] Y. N. Singh, "Pharmacology and Toxicology of Kava and Kavalactones," in Y. N. Singh, ed., *Kava from Ethnology to Pharmacology* (Boca Raton, FL: CRC Press, 2004).

[10] J. M. Mathews, A. S. Etheridge, J. L. Valentine, S. R. Black, D. P. Coleman, P. Patek, J. So, and L. T. Burka, "Pharmacokinetics and disposition of the kavalactone kawain: interaction with kava extract and kavalactones in vivo and in vitro," *Drug Metabolism and Disposition* 33 (2005): 1555-1563.

[11] J. Keledjian, P. H. Duffield, D. D. Jamieson, R. O. Lidgard, and A. M. Duffield, "Uptake into mouse brain of four compounds present in the psychoactive beverage kava," *Journal of Pharmaceutical Science* 77 (1988): 1003-1006.

[12] C. Backhauss and J. Krieglstein, "Extract of kava (Piper methysticum) and its methysticin constituents protect brain tissue against ischemic damage in rodents," *European Journal of Pharmacology* 215 (1992): 265-269.

[13] S. S. Baum, R. Hill, and H. Rommelspacher, "Effect of kava extract and individual kavapyrones on neurotransmitter levels in the nucleus accumbens of rats," *Progress in Neuropsychopharmacology and Biological Psychiatry* 22 (1998): 1105-1120.

[14] A. Jussofie, A. Schmiz, and C. Hiemke, "Kavapyrone enriched extract from Piper methysticum as modulator of the GABA binding site in different regions of rat brain," *Psychopharmacology (Berlin)* 116 (1994): 469-474.

[15] H. Grunze, J. Langosch, K. Schirrmacher, D. Bingmann, J. Von Wegerer, and J. Walden, "Kava pyrones exert effects on neuronal transmission and transmembraneous cation currents similar to established mood stabilizers—a review," *Progress in Neuropsychopharmacology and Biological Psychiatry* 25 (2001): 1555-1570.

[16] E. I. Magura, M. V. Kopanitsa, J. Gleich, T. Peters, and O. A. Krishtal, "Kava extract ingredients, (+)-methysticin and (+/-)-kavain inhibit voltage-operated Na(+)-channels in rat CA1 hippocampal neurons," *Neuroscience* 81 (1997): 345-351.

[17] U. Seitz, A. Schule, and J. Gleitz, "[$^{3}$H]-Monoamine uptake inhibition properties of kavapyrones," *Planta Medica* 63 (1997): 548-549.

[18] R. Uebelhack, L. Franke, and H. J. Schewe, "Inhibition of platelet MAO-B by kava pyrone-enriched extract from Piper methysticum Forster (kava-kava)," *Pharmacopsychiatry* 31 (1998): 187-192.

[19] A. A. Shaik, D. L. Hermanson, and C. Xing, "Identification of methysticin as a potent and non-toxic NF-kappaB inhibitor from kava, potentially responsible for kava's chemopreventive activity," *Bioorganic, Medicinal and Chemical Letters* 19 (2009): 5732-5736.

[20] R. J. Boerner, H. Sommer, W. Berger, U. Kuhn, U. Schmidt, and M. Mannel, "Kava-Kava extract LI 150 is as effective as opipramol and buspirone in generalized anxiety disorder—an 8-week randomized double-blind, multi-centre clinical trial in 129 out-patients," *Phytomedicine* 10, Suppl. 4 (2003): 38-49.

[21] J. Sarris, D. J. Kavanagh, G. Byrne, K. M. Bone, J. Adams, and G. Deed, "The Kava Anxiety Depression Spectrum Study (KADSS): a randomized, placebo-controlled crossover trial using an aqueous extract of Piper methysticum," *Psychopharmacology (Berlin)* 205 (2009): 399-407.

[22] F. P. Geier and T. Konstantinowicz, "Kava treatment in patients with anxiety," *Phytotherapy Research* 18 (2004): 297-300.

[23] S. Lehrl, "Clinical efficacy of kava extract WS1490 in sleep disturbances associated with anxiety disorders. Results of a multicenter, randomized, placebo-controlled, double-blind clinical trial," *Journal of Affective Disorders* 78 (2004): 101-110.

[24] S. Witte, D. Loew, and W. Gaus, "Meta-analysis of the efficacy of the acetonic kava-kava extract WS1490 in patients with non-psychotic anxiety disorders," *Phytotherapy Research* 19 (2005): 183-188.

[25] J. Sarris, E. La Porte, and I. Schweitzer, "Kava: a comprehensive review of efficacy, safety and psychopharmacology," *Australian and New Zealand Journal of Psychiatry* 45 (2011): 27-35.

[26] A. Rowe, L. Y. Zhang, and I. Ramzan, "Toxicokinetics of kava," *Advances in Pharmacological Sciences* (2011) e pub, DOI: 10.1155/2011/326724

[27] R. Teschke, S. X. Qiu, and V. Lebot, "Herbal hepatotoxicity by kava: Update on pipermethystine, flavokavain B, and mould hepatotoxin as primarily assumed culprits," *Digestive and Liver Disease* 43 (2011): 676-681.

[28] K. Dragull, W. Y. Yoshida, and C. S. Tang, "Piperidine alkaloids from Piper methysticum," *Phytochemistry* 63 (2003): 193-198.

[29] M. Lechtenberg, B. Quandt, M. Schmidt, and A. Nahrstedt, "Is the alkaloid pipermethystine connected with the claimed liver toxicity of Kava products?" *Pharmazie* 63 (2008): 71-74.

[30] S. T. Lim, K. Dragull, C. S. Tang, H. C. Bittenbender, J. T. Efird, and P. V. Nerurkar, "Effects of kava alkaloid, pipermethystine, and kavalactones on oxidative stress and cytochrome P450 in F-344 rats," *Toxicological Sciences* 97 (2007): 214-221.

[31] P. Zhou, S. Gross, J. H. Liu, B. Y. Yu, L. L. Feng, J. Nolta, V. Sharma, D. Piwnica-Worms, and S. X. Qiu, "Flavokawain B, the hepatoxic constituent from kava root induces GSH-sensitive oxidative stress through modulation of IKK/NF-kappaB and MAPK signaling pathways," *FASEB Journal* 24 (2010): 4722-4732.

[32] X. W. Chen, E. S. Serag, K. B. Sneed, and S. F. Zhou, "Herbal bioactivation, molecular targets and the toxicity relevance," *Chemico-Biological Interactions* 192 (2011): 161-176.

[33] X. Yang and W. F. Salminen, "Kava extract, an herbal alternative for anxiety relief, potentiates acetaminophen-induced cytotoxicity in rat hepatic cells," *Phytomedicine* 18 (2011): 592-600.

[34] Y. Ma, K. Sachdeva, J. Liu, M. Ford, D. Yang, I. A. Khan, C. O. Chichester, and B. Yan, "Desmethoxyyangonin and dihydromethysticin are two major pharmacological kavalactones with marked activity on the induction of CYP3A23," *Drug Metabolism and Disposition* 32 (2004): 1317-1324.

[35] B. J. Gurley, S. F. Gardner, M. A. Hubbard, D. K. Williams, W. B. Gentry, I. A. Khan, and A. Shah, "In vivo effects of goldenseal, kava kava, black cohosh, and valerian on human cytochrome P450 1A2, 2D6, 2E1 and 3A4/5 phenotypes," *Clinical Pharmacology and Therapeutics* 77 (2005): 415-426.

[36] J. Sarris, R. Teschke, C. Stough, A. Scholey, and I. Schweitzer, "Re-introduction of kava (Piper methysticum) to the EU: is there a way forward?" *Planta Medica* 77 (2011): 107-110.

## Chapter 10

[1] S. Norton, "The pharmacology of Mithridatum: a 2000 year-old remedy," *Molecular Interventions,* 6 (2006): 60-66.

[2] G. Andrea, *Libro de i Secretti con Ricette,* unpublished manuscript dated 1562, S. Norton, trans. (Lawrence, KS: The Kenneth Spencer Research Library, University of Kansas, trans. 2009).

[3] N. Culpeper, *Culpeper's Complete Herbal & English Physician,* 1826 edition (Reprint: Leicester, UK: Magna Books, 1992).

[4] J. Gerard, *The Herbal or General History of Plants,* 1633 edition (Reprint: New York: Dover Publications, Inc., 1975).

[5] M. Grieve, *A Modern Herbal* (New York: Barnes and Noble, Inc., 1996).

[6] H. M. Cavanagh and J. M. Wilkinson, "Biological activities of lavender essential oil," *Phytotherapy Research* 16 (2002): 301-308.

[7] Y. Guitton, F. Nicole, S. Moja, N. Valot, S. Legrand, F. Jullien, and L. Legendre, "Differential accumulation of volatile terpene and terpene synthase mRNAs during lavender (Lavandula angustifolia and L. x intermedia) inflorescence development," *Physiologia Planta* 138 (2010): 150-163.

[8] G. Woronuk, Z. Demissie, M. Rheault, and S. Mahmoud, "Biosynthesis and therapeutic properties of Lavandula essential oil constituents." *Planta Medica* 77 (2011): 7-15.

[9] J. B. Harborne and C. A. Williams, "Phytochemistry of the Genus Lavandula," in M. Lis-Balchin, ed., *Lavender: The Genus Lavandula* (New York: Taylor and Francis, 2002).

[10] A. Desantiles, K. Biswas, A. Lane, A. Boeckelmann, and S. S. Mahmoud, "Suppression of linalool acetate production in Lavandula x. intermedia," *Natural Products Communications* 4 (2009): 1533-1566.

[11] K. Cal, "Skin penetration of terpenes from essential oils and topical vehicles," *Planta Medica* 72 (2006): 311-316.

[12] K. Cal and M. Sznitowska, "Cutaneous absorption and elimination of three acyclic terpenes—in vitro studies," *Journal of Control Release* 93 (2003): 369-376.

[13] E. Heuberger, S. Redhammer, and G. Buchbauer, "Transdermal absorption of (-)-linalool induces autonomic deactivation but has no impact on ratings of well-being in humans," *Neuropsychopharmacology* 29 (2004): 1925-1932.

[14] K. Cal and M. Krzyzaniak, "*Stratum corneum* absorption and retention of linalool and terpine-4-ol applied as gel or oily solution in humans," *Journal of Dermatological Science* 42 (2006): 265-267.

[15] W. Jager, G. Buchbauer, L. Jirovetz, and M. Fritzer, "Percutaneous absorption of lavender oil from a massage oil," *Journal of the Society of Cosmetology Chemistry* 43 (1992): 49-54.

[16] R. Meesters, M. Duisken, and J. Hollender, "Study on the cytochrome P450-mediated oxidative metabolism of the terpene alcohol linalool: indication of biological epoxidation," *Xenobiotica* 37 (2007): 604-617.

[17] A. Chadha and K. Madyastha, "Metabolism of geraniol and linalool in the rat and effects on liver and lung microsomal enzymes," *Xenobiotica* 14 (1984): 365-374.

[18] A. Jori, A. Bianchetti, and P. Prestini, "Effect of essential oils on drug metabolism," *Biochemical Pharmacology* 18 (1969): 2081-2085.

[19] D. Parke and H. Rahman, "The effects of some terpenoids and other dietary nutrients on hepatic drug-metabolizing enzymes," *Biochemical Journal* 113 (1969): 12.

[20] G. Buchbauer, L. Jirovetz, W. Jager, H. Dietrich, and C. Plank, "Aromatherapy: evidence for sedative effects of the essential oil of lavender after inhalation," *Zeitschrift fur Naturforschung* 46 (1991): 1067-1072.

[21] V. M. Linck, A. L. da Silva, M. Figueiro, A. L. Piato, A. P. Herrmann, F. Dupont Birck, E. B. Caramao, D. S. Nunes, P. R. Moreno, and E. Elisabetsky, "Inhaled linalool-induced sedation in mice," *Phytomedicine* 16 (2009): 303-307.

[22] H. W. Tsang and T. Y. Ho, "A systematic review on the anxiolytic effects of aromatherapy on rodents under experimentally induced anxiety models," *Reviews in the Neurosciences* 21(2010): 141-152.

[23] T. Umezu, K. Nagano, H. Ito, K. Kosakai, M. Sakanuva, and M. Morita, "Anticonflict effects of lavender oil and identification of its active constituents," *Pharmacology, Biochemistry and Behavior* 85 (2006): 713-721.

[24] F. Souto-Maior, F. de Carvalho, L. de Morais, S. Netto, D. de Sousa, and R. de Almeida, "Anxiolytic-like effects of inhaled linalool oxide in experimental mouse anxiety models," *Pharmacology, Biochemistry and Behavior* 100 (2011): 259-263.

[25] V. Linck, A. da Silva, M. Figueiro, E. Caramao, P. Moreno, and E. Elisabetsky, "Effects of inhaled Linalool in anxiety, social interaction and aggressive behavior in mice," *Phytomedicine* 17 (2010): 679-683.

[26] D. Shaw, K. Norwood, and J. Leslie, "Chlordiazepoxide and lavender oil alter unconditioned anxiety-induced c-fos expression in the rat brain," *Behavioral Brain Research* 224 (2011): 1-7.

[27] J. Shen, A. Niijima, M. Tanida, Y, Horii, K. Maeda, and K. Nagai, "Olfactory stimulation with scent of lavender oil affects autonomic nerves, lipolysis and appetite in rats," *Neuroscience Letters* 383 (2005): 188-193.

[28] K. Yamada, Y. Mimakii, and Y. Sashida, "Effect of inhaling the vapor of Lavandula burnatii super-derived essential oil and linalool on plasma adrenocorticotropic hormone (ACTH), catecholamine and gonadotropin levels in experimental menopausal female rats," *Biological and Pharmaceutical Bulletin* 28 (2005): 1378-1379.

[29] U. Stockhorst and R. Pietrowsky, "Olfactory perception, communication, and the nose-to-brain pathway," *Physiology and Behavior* 83 (2004): 3-11.

[30] M. S. Kashani, M. R. Tavirani, S. A. Tolaei, and M. Salami, "Aqueous extract of lavender (Lavandula angustifolia) improves the spatial performance of a rat model of Alzheimer's disease," *Neuroscience Bulletin* 27 (2011): 99-106.

[31] B. J. Ding, W. W. Ma, L. L. He, X. Zhou, L. H. Yuan, H. L. Yu, J. F. Feng, and R. Xiao, "Soybean isoflavone alleviates beta-amyloid 1-42 induced inflammatory response to improve learning and memory by down regulation of Toll-like receptor 4 expression and nuclear factor KB activity in rats," *International Journal of Developmental Neuroscience* 29 (2011): 537-542.

[32] A. T. Peana, P. S. D'Aquila, F. Panin, G. Serra, P. Pippia, and M. D. Moretti, "Anti-inflammatory activity of linalool and linalyl acetate constituents of essential oils," *Phytomedicine* 9 (2002): 721-726.

[33] V. Coelho, J. Gianesini, R. Von Borowski, L. Mazzardo-Martins, D. Martins, J. Picada, A. Santos, L. Brum, and P. Pereira, "(-)-Linalool, a naturally occurring monoterpene compound, impairs memory acquisition in the object recognition task, inhibitory avoidance test and habituation to a novel environment in rats," *Phytomedicine* 18 (2011): 896-901.

[34] D. de Sousa, F. Nobrega, C. Santos, and R. de Almeida, "Anticonvulsant activity of the linalool enantiomers and racemate: investigation of chiral influence," *Natural Product Communications* 5 (2010): 1847-1851.

[35] T. Sakurada, H. Kuwahata, S. Katsuyama, T. Komatsu, L. Morrone, M. Corasaniti, G. Bagetta, and S. Sakurada, "Intraplanar injection of bergamot essential oil into the mouse hindpaw: effects on capsaicin-induced nociceptive behaviors," *International Review of Neurobiology* 85 (2009): 237-248.

[36] E. Barocelli, F. Calcina, M. Chiavarini, M. Impicciatore, R. Bruni, A. Bianchi, and V. Ballabeni, "Antinociceptive and gastroprotective effects of inhaled and orally administered Lavandula hybrid Reverchon 'Grosso' essential oil," *Life Sciences* 76 (2004): 213-223.

[37] A. Peana, P. D'Aquila, M. Chessa, M. Moretti, G. Serra, and P. Pippia, "(-)-Linalool produces antinociception in two experimental models of pain," *European Journal of Pharmacology* 460 (2003): 37-41.

[38] P. Batista, M. Werner, E. Oliveira, L. Burgos, P. Pereira, L. Brum, G. Story, and A. Santos, "The antinociceptive effect of (-)-linalool in models of chronic inflammatory and neuropathic hypersensitivity in mice," *Journal of Pain* 11 (2010): 1222-1229.

[39] A. Penana, M. De Montis, S. Sechi, G. Sircana, P. D'Aquila, and P. Pippia, "Effects of (-)-linalool in the acute hyperalgesia induced by carrageenan, L-glutamate and prostaglandin E2," *European Journal of Pharmacology* 497 (2004): 279-284.

[40] L. Berliocchi, R. Russo, A. Levato, V. Fratto, G. Bagetta, S. Sakurada, T. Sakurada, N. Mercuri, and M. Corasaniti, "(-)-Linalool attenuates allodynia in neuropathic pain induced by spinal nerve ligation in c57/bl6 mice," *International Review of Neurobiology* 85 (2009): 221-235.

[41] I. Zalachoras, A. Kagiava, D. Vokuo, and G. Theophilidis, "Assessing the local anesthetic effect of five essential oil constituents," *Planta Medica* 76 (2010): 1647-1653.

[42] C. Ghelardini, N. Galeotti, G. Salvatore, and G. Mazzanti, "Local anesthetic activity of the essential oil of Lavandula angustifolia," *Planta Medica* 65 (1999): 700-703.

[43] D. Trombetta, F. Castelli, M. G. Sarpietro, V. Venuti, M. Cristani, C. Daniele, A. Saija, G. Mazzamti, and G. Bisignano, "Mechanisms of antibacterial action of 3 monoterpenes," *Antimicrobial Agents and Chemotherapy,* 49 (2005): 2474-2478

[44] R. Koto, M. Imamura, C. Watanabe, S. Obayashi, M. Shiraishi, Y. Sasaki, and H. Azuma, "Linalyl acetate as a major ingredient of lavender essential oil relaxes the rabbit vascular smooth muscle through dephosphorylation of myosin light chain," *Journal of Cardiovascular Pharmacology* 48 (2006): 850-856.

[45] S. Cho, H. Jun, J. Lee, Y. Jia, K. Kim, and S. Lee, "Linalool reduces the expression of 3-hydroxy-3-methylglutaryl CoA reductase via sterol regulatory element binding protein-2- and ubiquitin-dependent mechanisms," *FEBS Letters* 585 (2011): 3289-3296.

[46] S. Kim, H. Kim, J. Yeo, S. Hong, J. Lee, and Y. Jeon, "The effect of lavender oil on stress, bispectral index values, and needle insertion pain in volunteers," *Journal of Alternative and Complimentary Medicine* 17 (2011): 823-826.

[47] M. Kritsidima, T. Newton, and K. Asimakopoulou, "The effects of lavender scent on dental patient anxiety levels: a cluster randomized-controlled trial," *Community Dentistry and Oral Epidemiology* 38 (2010): 83-87.

[48] L. Muzzarelli, M. Force, and M. Sebold, "Aromatherapy and reducing preprocedural anxiety: A controlled prospective study," *Gastroenterology Nursing* 29 (2006): 466-471.

[49] K. Kuroda, N. Inoue, Y. Ito, K. Kubota, A. Sugimoto, T. Kakuda, and T. Fushiki, "Sedative effects of the jasmine tea odor and (R)-(-)-linalool, one of its major odor components, on autonomic nerve activity and mood stress," *European Journal of Applied Physiology* 95 (2005): 107-114.

[50] N. Goel, H. Kim, and R. Lao, "An olfactory stimulus modifies nighttime sleep in young men and women," *Chronobiology International* 22 (2005): 889-904.

[51] C. Holmes, V. Hopkins, C. Hensford, V. MacLaughlin, D. Wilkinson, and H. Rosenvinge, "Lavender oil as a treatment for agitated behaviour in severe dementia: a placebo controlled study," *International Journal of Geriatric Psychiatry* 17 (2002): 305-308.

[52] D. Jimbo, Y. Kimura, M. Taniguchi, M. Inoue, and K. Urakami, "Effect of aromatherapy on patients with Alzheimer's disease," *Psychogeriatrics* 9 (2009): 173-179.

[53] P. W. Lin, W. C. Chan, B. F. Ng, and L. C. Lam, "Efficacy of aromatherapy (Lavandula angustifolia) as an intervention for agitated behaviours in Chinese older persons with dementia: a cross-over randomized trial," *International Journal of Geriatric Psychiatry* 22 (2007): 405-410.

[54] S. Kasper, M. Gastpar, W. Muller, H. Volz, H. Moller, A. Dienel, and S. Schlafke, "Efficacy and safety of silexan, a new, orally administered lavender oil preparation in subthreshold anxiety disorder—evidence from clinical trials," *Wiener Medizinische Wochenschrift* 160 (2010): 547-556.

[55] H. Woelk and S. Schlafke, "A multi-center, double-blind, randomized study of the Lavender oil preparation Silexan in comparison to Lorazepam for generalized anxiety disorder," *Phytomedicine* 17 (2010): 94-99.

[56] S. Kasper, M. Gastpar, W. E Muller, H. P. Volz, H. J. Moller, A. Dienel, and S. Schlafke, "Silexan, an orally administered Lavandula oil preparation, is effective in the treatment of 'subsyndromal' anxiety disorder: a randomized, double-blind, placebo controlled trial," *International Clinical Psychopharmacology* 25 (2010): 277-287.

[57] S. J. Hossain, H. Aoshima, H. Koda, and Y. Kiso, "Fragrances in oolong tea that enhance the response of GABAA receptors," *Bioscience, Biotechnology and Biochemistry* 68 (2004): 1842-1848.

[58] S. Hossain, K. Hamamoto, H. Aoshima, and Y. Hara, "Effects of tea components on the response of GABA(A) receptors expressed in Xenopus Oocytes," *Journal of Agricultural and Food Chemistry* 50 (2002): 3954-3960.

[59] M. Cline, J. Taylor, J. Flores, S. Bracken, S. McCall, and T. Ceremuga, "Investigation of the anxiolytic effects of linalool, a lavender extract, in the male Sprague-Dawley rat," *AANA Journal* 76 (2008): 47-52.

[60] L. Brum, E. Elisabetsky, and D. Souza, "Effects of linalool on [$^3$H] MK801 and [$^3$H] muscimol binding in mouse cortical membranes," *Phytotherapy Research* 15 (2001): 422-425.

[61] L. Brum, T. Emanuelli, D. Souza, and E. Elisabetsky, "Effects of linalool on glutamate release and uptake in mouse cortical synaptosomes," *Neurochemical Research* 26 (2001): 191-194.

[62] P. Batista, M. Werner, E. Oliveira, L. Burgos, P. Pereira, L. Brum, and A. Santos, "Evidence for the involvement of ionotropic glutamatergic receptors on the antinociceptive effect (-)-linalool in mice," *Neuroscience Letters* 440 (2008): 299-303.

[63] E. Elisabetsky, L. Brum, and D. Souza, "Anticonvulsant properties of linalool in glutamate-related seizure models," *Phytomedicine* 6 (1999): 107-113.

[64] S. Dohi, M. Terasaki, and M. Makino, "Acetylcholinesterase inhibitory activity and chemical composition of commercial essential oils," *Journal of Agricultural Food Chemistry* 57 (2009): 4313-4318.

[65] J. Leal-Cardoso, K. da Silva-Alves, F. Ferreira-da-Silva, T. dos Santos-Nascimento, H. Joca, F. de Macedo, P. de Albuquerque-Neto, P. Magalhaes, S. Lahlou, J. Cruz, and R. Barbosa, "Linalool blocks excitability in peripheral nerves and voltage-dependent $Na^+$ current in dissociated dorsal root ganglia neurons," *European Journal of Pharmacology* 645 (2010): 86-93.

[66] L. Re, S. Barocci, S. Sonnino, A. Mencarelli, C. Vivani, G. Paolucci, A. Scarpantonio, L. Rinaldi, and E. Mosca, "Linalool modifies the nicotinic receptor-ion channel kinetics at the mouse neuromuscular junction," *Pharmacological Research* 42 (2000): 177-182.

[67] D. Henley, N. Lipson, K. Korach, and C. Bloch, "Prepubertal gynecomastia linked to lavender and tea tree oils," *New England Journal of Medicine* 356 (2007): 479-485.

[68] C. Da Porto and D. Decorti, "Analysis of the volatile compounds of flowers and essential oils from Lavandula angustifolia cultivated in Northeastern Italy by headspace solid-phase microextraction coupled to gas chromatography-mass spectrometry," *Planta Medica* 74 (2008): 182-187.

[69] D. Buckley, "Allergy to oxidized linalool in the UK," *Contact Dermatitis* 64 (2011): 240-241.

## Chapter 11

[1] Q. Wang, S. R. Manchester, and D. L. Dilcher, "Fruits and foliage of Pueraria (Leguminosae, Papilionoideae) from the Neogene of Eurasia and their biogeographic implications," *American Journal of Botany* 97 (2010): 1982-1998.

[2] L. J. G. Maesen, "Pueraria: Botanical Characteristics," in W. M. Keung, ed., *Pueraria, the genus Pueraria* (New York: Taylor & Francis, 2002).

[3] S-C Li, *Chinese Medicinal Herbs*, F. Porter Smith and G. A. Stuart, trans., (Mineola, NY: Dover Publications, Inc., 2003).

[4] H.-M. Chang and P. P.-H. But, eds., *Pharmacology and Applications of Chinese Materia Medica* (Singapore: World Scientific Publishing Co., 1987).

[5] M. P. Arolfo, D. H. Overstreet, L. Yao, P. Fan, A. J. Lawrence, G. Tao, W. M. Keung, B. L. Vallee, M. F. Olive, J. T. Gass, E. Rubin, H. Anni, C. W. Hodge, J. Besheer, J. Zablocki, K. Leung, B. K. Blackburn, L. G. Lange, and I. Diamond, "Suppression of heavy drinking and alcohol seeking by a selective ALDH-2 inhibitor," *Alcoholism, Clinical and Experimental Research* 33 (2009): 1935-1944.

[6] D. H. Overstreet, J. E. Kralic, A. L. Morrow, Z. Z. Ma, Y. W. Zhang, and D. Y. Lee, "NPI-031G (puerarin) reduces anxiogenic effects of alcohol withdrawal or benzodiazepine inverse or 5-HT2C agonists," *Pharmacology, Biochemistry and Behavior* 75 (2003): 619-625.

[7] H. Rong, D. De Keukeleire, and L. De Cooman, "Chemical Constituents of *Pueraria* Plants: Identification and Methods of Analysis," in W. M. Keung, ed., *Pueraria, the genus Pueraria* (NewYork: Taylor & Francis, 2002).

[8] C. Zhao, H. Y. Chan, D. Yuan, Y. Liang, T. Y. Lau, and F. T. Chau, "Rapid simultaneous determination of major isoflavones of Pueraria lobata and discriminative analysis of its geographical origins by principal component analysis," *Phytochemical Analysis* 22 (2011): 503-508.

[9] L. Bebrevska, K. Foubert, N. Hermans, S. Chatterjee, E. Van Mark, G. De Meyer, A. Vlietinek, L. Pieters, and S. Apers, "In vivo antioxidative activity of a quantified Pueraria lobata root extract," *Journal of Ethnopharmacology* 127 (2010): 112-117.

[10] C. Sibao, Y. Dajian, C. Shilin, X. Hongx, and A. S. Chan, "Seasonal variations in the isoflavonoids of radix Puerariae," *Phytochemical Analysis* 18 (2007): 245-250.

[11] J. Kaludjerovic, A. A. Franke, and W. E. Ward, "Circulating isoflavonoid levels in CD-1 mice: effect of oral versus subcutaneous delivery and frequency of administration," *Journal of Nutritional Biochemistry* e-pub 21658927.

[12] K. D. Setchell, N. M. Brown, P. Desai, L. Zimmer-Nechemias, B. E. Wolfe, W. T. Brashear, A. S. Kirschner, A. Cassidy, and J. E. Heubi, "Bioavailability of pure isoflavones in healthy humans and analysis of commercial soy isoflavone supplements," *Journal of Nutrition* 131, Suppl. 4 (2001): 1362S-1375S.

[13] I. L. Nielsen and G. Williamson, "Review of the factors affecting bioavailability of soy isoflavones in humans," *Nutrition and Cancer* 57 (2007): 1-10.

[14] M. K. Piskula, J. Yamakoshi, and Y. Iwai, "Daidzein and genistein but not their glucosides are absorbed from the rat stomach," *FEBS Letters* 447 (1999): 287-291.

[15] K. Murota, S. Shimizu, S. Miyamoto, T. Izumi, A. Obata, M. Kikuchi, and J. Terao, "Unique uptake and transport of isoflavone aglycones by human intestinal caco-2 cells: comparison of isoflavonoids and flavonoids," *Journal of Nutrition* 132 (2002): 1956-1961.

16 K. D. Setchell, N. M. Brown, L. Zimmer-Nechemias, W. T. Brashear, B. E. Wolfe, A. S. Kirschner, and J. E. Heubi, "Evidence for lack of absorption of soy isoflavone glycosides in humans, supporting the crucial role of intestinal metabolism for bioavailability," *American Journal of Clinical Nutrition* 76 (2002): 447-453.

17 J-S Jin, T. Nishihata, N. Kakiuchi, and M. Hattori, "Biotransformation of C-glucosylisoflavone puerarin to estrogenic (3S)-equol in co-culture of two human intestinal bacteria," *Biological and Pharmaceutical Bulletin* 31 (2008): 1621-1625.

18 J. K. Prasain, N. Peng, E. Acosta, R. Moore, A. Arabshahi, E. Meezan, S. Barnes, and J. M. Wyss, "Pharmacokinetic study of puerarin in rat serum by liquid chromatography tandem mass spectrometry," *Biomedical Chromatography* 21 (2007): 410-414.

19 Y. Li, W. S. Pan, S. L. Chen, H. X. Xu, D. J. Yang, and A. S. Chan, "Pharmacokinetic, tissue distribution, and excretion of puerarin and puerarin-phospholipid complex in rats," *Drug Development and Industrial Pharmacy* 32 (2006): 413-422.

20 C. F. Luo, M. Yuan, M. S. Chen, S. M. Liu, L. Zhu, B. Y. Huang, X. W. Liu, and W. Xiong, "Pharmacokinetics, tissue distribution and relative bioavailability of puerarin solid lipid nanoparticles following oral administration," *International Journal of Pharmacology* 410 (2011): 138-144.

21 J. G. Mun, M. D. Grannan, P. J. Lachcik, A. Reppert, G. G. Yousef, R. B. Rogers, E. M. Janle, C. M. Weaver, and M. H. Lila, "In vivo metabolic tracking of 14C-radiolabelled isoflavones in kudzu (Pueraria lobata) and red clover (Trifolium pretense) extracts," *British Journal of Nutrition* 102 (2009): 1523-1530.

22 J. K. Prasain, N. Peng, R. Moore, A. Arabshahi, S. Barnes, and J. M. Wyss, "Tissue distribution of puerarin and conjugated metabolites in rats assessed by liquid chromatography-tandem mass spectrometry," *Phytomedicine* 16 (2009): 65-71.

23 J. K. Prasain, K. Jones, N. Brissie, R. Moore, J. M. Wyss, and S. Barnes, "Identification of puerarin and its metabolites in rats by liquid chromatography-tandem mass spectrometry," *Journal of Agricultural and Food Chemistry* 52 (2004): 3708-3712.

24 E. Sepehr, G. M. Cooke, P. Robertson, and G. S. Gilani, "Effect of glycosidation of isoflavones on their bioavailability and pharmacokinetics in aged male rats," *Molecular Nutrition and Food Research* 53, Suppl.1 (2009): S16-S26.

25 J. E. Shin, E. A. Bae, Y. C. Lee, J. Y. Ma, and D. H. Kim, "Estrogenic effect of main components kakkalide and tectoridin of Puerariae Flos and their metabolites," *Biological and Pharmaceutical Bulletin* 29 (2006): 1202-1206.

26 Y. Xiong, Y. Yang, J. Yang, H. Chai, Y. Li, J. Yang, Z. Jia, and Z. Wang, "Tectoridin, an isoflavone glycoside from the flower of Pueraria lobata, prevents acute ethanol-induced liver steatosis in mice," *Toxicology* 276 (2010): 64-72.

[27] L. Lu, Y. Liu, W. Zhu, J. Shi, Y. Liu, W. Ling, and T. R. Kosten, "Traditional medicine in the treatment of drug addiction," *American Journal of Drug and Alcohol Abuse* 35 (2009): 1-11.

[28] E. Benihabib, J. I. Baker, D. E. Keyler, and A. K. Singh, "Kudzu root extract suppresses voluntary alcohol intake and alcohol withdrawal symptoms in P rats receiving free access to water and alcohol," *Journal of Medicinal Food* 7 (2004): 168-179.

[29] A. H. Rezvani, D. H. Overstreet, M. Perfumi, and M. Massi, "Plant derivatives in the treatment of alcohol dependency," *Pharmacology, Biochemistry and Behavior* 75 (2003): 593-606.

[30] E. Benihabib, J. I. Baker, D. E. Keyler, and A. K. Singh, "Effects of purified puerarin on voluntary alcohol intake and alcohol withdrawal symptoms in P rats receiving free access to water and alcohol," *Journal of Medicinal Food* 7 (2004): 180-186.

[31] J. Shebek and J. P. Rindone, "A pilot study exploring the effect of kudzu root on the drinking habits of patients with chronic alcoholism," *Journal of Alternative and Complementary Medicine* 6 (2000): 45-48.

[32] S. E. Lukas, D. Penetar, J. Berko, L. Vicens, C. Palmer, G. Mallya, E. A. Macklin, and D. Y. Lee, "An extract of Chinese herbal root kudzu reduces alcohol drinking by heavy drinkers in a naturalistic setting," *Alcoholism, Clinical and Experimental Research* 29 (2005): 756-762.

[33] N. Rooke, D. J. Li, J. Li, and W. M. Keung, "The mitochondrial monoamine oxidase-aldehyde dehydrogenase pathway: a potential site of action of daidzin," *Journal of Medicinal Chemistry* 43 (2000): 4169-4179.

[34] S. Svensson, J-O Hoog, G. Schneider, and T. Sandalova, "Crystal structures of mouse class II alcohol dehydrogenase reveal determinants of substrate specificity and catalytic efficiency," *Journal of Molecular Biology* 302 (2000): 441-453.

[35] W. M. Keung and B. L. Vallee, "Daidzin and its antidipsotropic analogs inhibit serotonin and dopamine metabolism in isolated mitochondria," *Proceedings of the National Academy of Sciences USA* 95 (1998): 2198-2203.

[36] G. Y. Gao, D. J. Li, and W. M. Keung, "Synthesis of potential antidipsotropic isoflavones: inhibitors of the mitochondrial monoamine oxidase-aldehyde dehydrogenase pathway," *Journal of Medicinal Chemistry* 44 (2001): 3320-3328.

[37] G. Y. Gao and W. M. Keung, "Synthesis of daidzin analogues as potential agents for alcohol abuse," *Bioorganic and Medicinal Chemistry Letters* 11 (2003): 4069-4081.

[38] R. C. Lin and T. K. Li, "Effects of isoflavones on alcohol pharmacokinetics and alcohol-drinking behavior in rats," *American Journal of Clinical Nutrition* 68, Suppl.6 (1998): S1512-S1515.

[39] G. Xing, M. Dong, X. Li, Y. Zou, L. Fan, X. Wang, D. Cai, C. Li, L. Zhou, J. Liu, and Y. Niu, "Neuroprotective effects of puerarin against beta-amyloid-induced neurotoxicity in PC12 cells via a P13K-dependent signaling pathway," *Brain Research Bulletin* 85 (2011): 212-218.

[40] G. Zhu, X. Wang, Y. Chen, S. Yang, H. Cheng, N. Wang, and Q. Li, "Puerarin protects dopaminergic neurons against 6-hydroxydopamine neurotoxicity via inhibiting apoptosis and upregulating glia cell line-derived neurotropic factor in a rat model of Parkinson's disease," *Planta Medica* 76 (2010): 1820-1826.

[41] D. H. Kim, H. A. Jung, S. J. Park, J. M. Kim, S. Lee, J. S. Choi, J. H. Cheong, K. H. Ko, and J. H. Ryu, "The effects of daidzin and its aglycon, daidzein, on the scopolamine-induced memory impairment in male mice," *Archives of Pharmacal Research* 33 (2010): 1685-1690.

[42] L. Zhao, Q. Chen, and R. Diaz Brinton, "Neuroprotective and neurotrophic efficacy of phytoestrogens in cultured hippocampal neurons," *Experimental Biology and Medicine* 227 (2002): 509-519.

[43] R. Vera, M. Galisteo, I. C. Villar, M. Sanchez, A. Zarzuelo, F. Perez-Vizcaino, and J. Duarte, "Soy isoflavones improve endothelial function in spontaneously hypertensive rats in an estrogen-independent manner: role of nitric-oxide synthase, superoxide, and cyclooxygenase metabolites," *Journal of Pharmacology and Experimental Therapeutics* 314 (2005): 1300-1309.

[44] M Hamalainen, R. Nieminen, M. Z. Asmawi, P. Vuorela, H. Vapaatalo, and E. Moilanen, "Effects of flavonoids on prostaglandin E2 production and on COX-2 and mPGES-1 expressions in activated macrophages," *Planta Medica* 77 (2011): 1504-1511).

[45] J. F. Regal, D. G. Fraser, C. E. Weeks, and N. A. Greenberg, "Dietary phytoestrogens have anti-inflammatory activity in a guinea pig model of asthma," *Proceedings of the Society for Experimental Biology and Medicine* 223 (2000): 372-378.

[46] T. Shimazu, M. Inoue, S. Sasazuki, M. Iwasaki, N. Sawada, T. Yamaji, and S. Tsugane, "Plasma isoflavones and the risk of lung cancer in women: a nested case-control study in Japan," *Cancer Epidemiology, Biomarkers and Prevention* 20 (2011): 419-427.

[47] Y. Chen, C. Q. Xiao, Y. J. He, B. L. Chen, G. Wang, G. Zhou, W. Zhang, Z. R. Tan, S. Cao, L. P. Wang, and H. H. Zhou, "Genistein alters caffeine exposure in healthy female volunteers," *European Journal of Clinical Pharmacology* 67 (2011): 347-353.

[48] J. Zheng, B. Chen, B. Jiang, L. Zeng, Z. R. Tang, L. Fan, and H. H. Zhou, "The effects of puerarin on CYP2D6 and CYP1A2 activities in vivo," *Archives of Pharmacal Research* 33 (2010): 243-246.

[49] T. Valachovicova, V. Slivova, H. Bergman, J. Shuherk, and D. Sliva, "Soy isoflavones suppress invasiveness of breast cancer cells by the inhibition of NF-kappaB/AP-1-dependent and –independent pathways," *International Journal of Oncology* 25 (2004): 1389-1395.

[50] T. Okura, M. Ibe, K. Umegaki, K. Shinozuka, and S. Yamada, "Effects of dietary ingredients on function and expression of P-glycoprotein in human intestinal epithelial cells," *Biological and Pharmaceutical Bulletin* 33 (2010): 255-259.

[51] M. Claussnitzer, T. Skurk, H. Hauner, H. Daniel, and M. J. Rist, "Effect of flavonoids on basal and insulin-stimulated 2-deoxyglucose uptake in adipocytes," *Molecular Nutrition and Food Research* 55, Suppl.1 (2011): S26-S34.

[52] N. R. McGregor, "Pueraria lobata (kudzu root) hangover remedies and acetaldehyde-associated neoplasm risk," *Alcohol* 41 (2007): 469-478.

[53] C. I. Xie, R. C. Lin, V. Antony, L. Lumeng, T. K. Li, K. Mai, C. Liu, Q. D. Wang, Z. H. Zhao, and G. F. Wang, "Daidzin, an antioxidant isoflavonoid, decreases blood alcohol levels and shortens sleep time induced by ethanol intoxication," *Alcohol Clinical Experimental Research* 18 (1994): 1443-1447.

[54] T. T. Hien, H. G. Kim, E. H. Han, K. W. Kang, and H. G. Jeong, "Molecular mechanism of suppression of mDR1 by puerarin from Pueraria lobata via NF-kappaB pathway and cAMP-responsive element transcriptional activity-dependent up-regulation of AMP-activated protein kinase in breast cancer MFC-7/adr cells," *Molecular Nutrition and Food Research* 54 (2010): 918-928.

## Chapter 12

[1] Theophrastus, *Enquiry into Plants*, Book IV, A. F. Hort, trans. (Cambridge, MA: Harvard University Press, 1980).

[2] J. Gerard, *The Herbal or General History of Plants*, 1633 ed., revised by T. Johnson (New York: Dover Publications, Inc., 1975).

[3] J. Bastida and F. Viladomat, "Alkaloids of Narcissus," G. R. Hanks, ed., *Narcissus and Daffodil* (New York: Taylor and Francis, 2002).

[4] A. Lubbe, Y. H. Choi, P. Vreeburg, and R. Verpoorte, "Effect of fertilizers on galanthamine and metabolite profiles in Narcissus bulbs by 1H NMR," *Journal of Agricultural and Food Chemistry* 59 (2011): 3155-3161.

[5] Pliny, *Natural History*, Book XXI, W. H. Jones, trans. (Cambridge, MA: Harvard University Press, 1999).

[6] M. Mashovsky and R. Kruglikova-Lvova, "On the pharmacology of the new alkaloid galantamine," *Farmakologia Toxicologia* 14 (1951): 27-30.

[7] L. van Beijsterveldt, R. Geerts, T. Verhaeghe, B. Willems, W. Bode, K. Lavrijsen, and W. Meuldermans, "Pharmacokinetics and tissue distribution of galantamine and galantamine-related radioactivity after single intravenous and oral administration in the rat," *Arzneimittelforschung* 54 (2004): 85-94.

[8] U. Bickel, T. Thomsen, J. Fischer, W. Weber, and H. Kewitz, "Galanthamine: pharmacokinetics, tissue distribution and cholinesterase inhibition in brain of mice," *Neuropharmacology* 30 (1991): 447-454.

9 D. Mihailova, I. Yamboliev, Z. Zhivkova, J. Tencheva, and V. Jovovich, "Pharmacokinetics of galanthamine hydrobromide after single subcutaneous and oral dosage in humans," *Pharmacology* 39 (1989): 50-58.

10 U. Bickel, T. Thomsen, W. Weber, J. P. Fischer, R. Bachus, M. Nitz, and H. Kewitz, "Pharmacokinetics of galanthamine in humans and corresponding cholinesterase inhibition," *Clinical Pharmacology and Therapeutics* 50 (1991): 420-428.

11 F. Huang and Y. Fu, "A review of clinical pharmacokinetics and pharmacodynamics of galantamine, a reversible acetylcholinesterase inhibitor for the treatment of Alzheimer's disease, in healthy subjects and patients," *Current Clinical Pharmacology* 5 (2010): 115-124.

12 S. Kretzing, G. Abraham, B. Seiwert, F. R. Ungemach, U. Krugel, and R. Regenthal, "Dose-dependent emetic effects of the Amaryllidaceous alkaloid lycorine in beagle dogs," *Toxicon* 57 (2011): 117-124.

13 J. McNulty, J. Nair, M. Singh, D. Crankshaw, A. Holloway, and J. Bastida, "Cytochrome P450 3A4 inhibitory activity studies with the lycorine series of alkaloids," *Natural Product Communications* 5 (2010): 1195-2000.

14 A. Maelicke and E. Albuquerque, "Allosteric modulation of nicotinic acetylcholine receptors as a treatment strategy for Alzheimer's disease," *European Journal of Pharmacology* 393 (2000): 165-170.

15 A. Maelicke, A. Schrettenholz, M. Samochocki, M. Radina, and E. X. Albuquerque, "Allosterically potentiating ligands of nicotinic receptors as a treatment strategy for Alzheimer's disease," *Behavioural Brain Research* 113 (2000): 199-206.

16 M. Samochocki, A. Hoffle, A. Fehrenbacher, R. Jostock, J. Ludwig, C. Christner, M. Radina, M. Zerlin, C. Ullmer, E. F. Pereira, H. Lubbert, E. X. Albuquerque, and M. Maelilcke, "Galantamine is an allosterically potentiating ligand of neuronal nicotinic but not of muscarinic acetylcholine receptors," *Journal of Pharmacology and Experimental Therapeutics* 305 (2003): 1024-1036.

17 G. M. Bores, F. P. Huger, W. Petko, A. E. Mutlib, F. Camacho, D. K. Rush, D. E. Selk, V. Wolf, R. W. Kosley Jr., L. Davis, and H. M. Vargas, "Pharmacological evaluation of novel Alzheimer's disease therapeutics: acetylcholinesterase inhibitors related to galanthamine," *Journal of Pharmacology and Experimental Therapeutics* 277 (1996): 728-738.

18 J. Birks, "Cholinesterase inhibitors for Alzheimer's disease," *Cochrane Database of Systematic Reviews* (2006), Issue 1. Art. No.: CD005593. DOI: 10.1002/14651858.CD005593.

19 C. Loy and L. Schneider, "Galantamine for Alzheimer's disease and mild cognitive impairment," *Cochrane Database of Systematic Reviews* (2006), Issue 1. Art. No.: CD001747. DOI:10.1002/14651858.CD001747.pub3.

20 S. Kavanagh, M. Gaudig, B. Van Baelen, M. Adami, A. Delgado, C. Guzman, E. Jedenius, and B. Schauble, "Galantamine and behavior in Alzheimer's disease: analysis of four trials," *Acta Neurologica Scandinavica* 124 (2011): 302-308.

[21] D. Wilkinson and J. Murray, "Galantamine: a randomized, double-blind, dose comparison in patients with Alzheimer's disease," *International Journal of Geriatric Psychiatry* 16 (2001): 852-857.

[22] B. Fulton and P. Benfield, "Galanthamine," *Drugs and Aging* 9 (1996): 60-65.

[23] M. Gaudig, U. Richarz, J. Han, B. V. Baelen, and B. Schauble, "Effect of galantamine in Alzheimer's disease: double-blind withdrawal studies evaluating sustained versus interrupted treatment," *Current Alzheimer Research* 8 (2011): 771-780.

[24] C. Chianella, D. Gragnaniello, P. Maisano Delser, M. F. Visentini, E. Sette, M. R. Tola, G. Barbujani, and S. Fuselli, "BCHE and CYP2D6 genetic variation in Alzheimer's disease patients treated with cholinesterase inhibitors," *European Journal of Clinical Pharmacology* 67 (2011): 1147-1157).

[25] D. A. Cozanitis, T. Friedmann, and S. Furst, "Study of the analgesic effects of galanthamine, a cholinesterase inhibitor," *Archives of International Pharmacodynamics and Therapeutics* 266 (1983): 229-238.

[26] S. K. Satapathy, M. Ochani, M. Dancho, L. K. Hudson, M. Rosas-Ballina, S. I. Valdes-Ferrer, P. S. Olofsson, Y. T. Harris, J. Roth, S. Chavan, K. J. Tracey, and V. A. Pavlov, "Galantamine alleviates obesity, inflammation and other obesity-associated complications in high-fat diet-fed mice," *Molecular Medicine* 17 (2011): 499-606.

[27] Y. Yamazaki and Y. Kawano, "Inhibitory effects of herbal alkaloids on the tumor necrosis factor-alpha and nitric oxide production in lipopolysaccharide-stimulated RAW264 macrophages," *Chemical and Pharmaceutical Bulletin* 59 (2011): 388-391.

[28] J. Kang, Y. Zhang, X. Cao, J. Fan, G. Li, Q. Wang, Y. Diao, Z. Zhao, L. Luo, and Z. Yin, "Lycorine inhibits lipopolysaccharide-induced iNOS and COX-2 upregulation in RAW264.7 cells through suppressing P38 and STATs activation and increases the survival rate of mice after LPS challenge," *International Immunopharmacology* 12 (2012): 249-256.

[29] D. Lamoral-Theys, A. Andolfi, G. Van Goietsenoven, A. Cimmino, B. Le Calve, N. Wauthoz, V. Megalizzi, T. Gras, C. Bruyere, J. Dubois, V. Maltheiu, A. Kornienko, R. Kiss, and A. Evidente, "Lycorine, the main phenanthradine Amaryllidaceae alkaloid, exhibits significant antitumor activity in cancer cells that display resistance to proapoptotic stimuli: an investigation of structure-activity relationship and mechanistic insight," *Journal of Medicinal Chemistry* 52 (2009): 6244-6256.

[30] G. Van Goietsenoven, A. Andolfi, B. Lallemand, A. Cimmino, D. Lamoral-Theys, T. Gras, A. Abou-Donia, J. Dubois, F. Lefranc, V. Mathieu, A. Kornienko, R. Kiss, and A. Evidente, "Amaryllidaceae alkaloids belonging to different structural subgroups display activity against apoptosis-resistant cancer cells," *Journal of Natural Products* 73 (2010): 1223-1227.

[31] L. De Gara and F. Tommasi, "Further researches upon the inhibiting action of lycorine on ascorbic acid biosynthesis," *Bollettino della Societa Italiana di Biologia Sperimentale* 66 (1990): 953-960.

[32] R. Liso, G. Calabrese, M. B. Bitonti, and O. Arrigoni, "Relationship between ascorbic acid and cell division," *Experimental Cell Research* 150 (1984): 314-320.

[33] J. Liu, Y. Yang, Y. Xu, C. Ma, C. Qin, and L. Zhang, "Lycorine reduces mortality of human enterovirus 71-infected mice by inhibiting virus replication," *Virology Journal* 8 (2011): 483.

[34] S. Kretzing, G. Abraham, B. Seiwert, F. R. Ungemach, U. Krugel, and R. Regenthal, "In vivo assessment of antiemetic drugs and mechanism of lycorine-induced nausea and emesis," *Archives of Toxicology* 85 (2011): 1565-1573.

[35] J. J. Sramek, E. J. Frackiewicz, and N. R. Cutler, "Review of the acetylcholinesterase inhibitor galanthamine," *Expert Opinion on Investigational Drugs* 9 (2000): 2393-2402.

## Chapter 13

[1] J. E. Bare, *Wildflowers and Weeds of Kansas* (Lawrence, KS: Regents Press of Kansas, 1979).

[2] M. Wichtl, *Herbal Drugs and Phytopharmaceuticals*, N. G. Bisset, ed., English edition, (Boca Raton, FL: CRC Press, 1994).

[3] K. Appel, T. Rose, B. Fiebich, T. Kammler, C. Hoffmann, and G. Weiss, "Modulation of the gamma-aminobutyric (GABA) system by Passiflora incarnata L.," *Phytotherapy Research* 25 (2011): 838-843.

[4] R. Masteikova, J. Bernatoniene, R. Bernatoniene, and S. Velziene, "Antiradical activities of the extract of Passiflora incarnata," *Acta Poloniae Pharmaceutica* 65(2008): 577-583.

[5] S. Akhandzadeh, L. Kashani, M. Mabaseri, S. H. Hosseini, S. Nikzad, and M. Khani, "Passionflower in the treatment of opiates withdrawal: a double-blind randomized controlled trial," *Journal of Clinical Pharmacology and Therapeutics* 26 (2001): 369-373.

[6] K. Dhawan, S. Dhawan, and A. Sharma, "Passiflora: a review update," *Journal of Ethnopharmacology* 94 (2004): 1-23.

[7] S. D. Muller, S. B. Vasconcelos, M. Coelho, and M. W. Biavatti, "LC and UV determination of flavonoids from Passiflora alata medicinal extracts and leaves," *Journal of Pharmaceutical and Biomedical Analysis* 37 (2005): 399-403.

[8] Q. M. Li, H. van den Heuvel, O. Delorenzo, J. Carthout, L. A. Pieters, A. J. Vlietinck, and M. Claeys, "Mass spectral characterization of C-glycosidic flavonoids isolated from a medicinal plant (Passiflora incarnata)," *Journal of Chromatography* 562 (1991): 435-446.

[9] R. Soulimani, C. Younos, S. Jarmouni, D. Bousta, R. Misslin, and F. Mortier, "Behavioural effects of Passiflora incarnata L. and its indole alkaloid and flavonoid derivatives and maltol in the mouse," *Journal of Ethnopharmacology* 57 (1997): 11-20.

[10] K. Pereira, S. Botelho-Junior, D. Domingues, O. Marchado, A. Oliveira, K. Fernandes, H. Madureira, T. Pereira, and T. Jacinto, "Passion fruit flowers: Kunitz trypsin inhibitors and cystatin differentially accumulate in developing buds and floral tissues," *Phytochemistry* 72 (2011): 1955-1961.

[11] K. Dhawan, S. Kumar, and A. Sharma, "Anxiolytic activity of aerial and underground parts of Passiflora incarnata," *Fitoterapia* 72 (2001): 922-926.

[12] H. Wohlmuth, K. G. Penman, T. Pearson, and R. P. Lehmann, "Pharmacognosy and chemotypes of passionflower (Passiflora incarnata L.)," *Biological and Pharmaceutical Bulletin* 33 (2010): 1015-1018.

[13] O. Grundmann, C. Wahling, C. Staiger, and V. Butterweck, "Anxiolytic effects of a passion flower (Passiflora incarnata L.) extract in the elevated plus maze in mice," *Pharmazie* 64 (2009): 63-64.

[14] C. Sampath, M. Holbik, L. Krenn, and V. Butterweck, "Anxiolytic effects of fractions obtained from Passiflora incarnata L. in the elevated plus maze in mice," *Phytotherapy Research* 25 (2011): 789-795.

[15] U. Walle, A. Galijatovic, and T. Walle, "Transport of the flavonoid chrysin and its conjugated metabolites by the human intestinal cell line Caco-2," *Biochemical Pharmacology* 58 (1999): 431-438.

[16] Y. Zhang, X. Tie, B. Bao, X. Wu, and Y. Zhang, "Metabolism of flavone C-glucosides and p-coumaric acid from antioxidant of bamboo leves (AOB) in rats," *British Journal of Nutrition* 97 (2007): 484-494.

[17] L. Y. Ma, R. H. Liu, X. D. Xu, M. Q. Yu, Q. Zhang, and H. L. Liu, "The pharmacokinetics of C-glycosyl flavones of Hawthorn leaf flavonoids in rat after single dose administration," *Phytomedicine* 17 (2010): 640-645.

[18] M. Liang, W. Xu, W. Zhang, C. Zhang, R. Liu, Y. Shen, H. Li, X. Wang, X. Wang, Q. Pan, and C. Chen, "Quantitative LC/MS/MS method and in vivo pharmacokinetic studies of vitexin rhamnoside, a bioactive constituent on cardiovascular system from hawthorn," *Biomedical Chromatography* 21 (2007): 422-429.

[19] H. Cai, D. Boocock, W. Steward, and A. Gescher, "Tissue distribution in mice and metabolism in murine and human liver of apigenin and tricin, flavones with putative cancer chemopreventive properties," *Cancer Chemotherapy and Pharmacology* 60 (2007): 257-266.

[20] L. Hanske, G. Loh, S. Sczesny, M. Blaut, and A. Braune, "The bioavailability of apigenin-7-glycoside is influenced by human intestinal microbiota in rats," *Journal of Nutrition* 139 (2009): 1095-1102.

[21] A. Gradolatto, J. Basly, R. Berges, C. Teyssier, M. Chagnon, M. Siess, and M. Canivenc-Lavier, "Pharmacokinetics and metabolism of apigenin in female and male rats after a single oral administration," *Drug Metabolism and Disposition* 33 (2005): 49-54.

[22] H. Meyer, A. Bolarinwa, G. Wolfram, and J. Linseisen, "Bioavailability of apigenin from apiin-rich parsley in humans," *Annals of Nutrition and Metabolism* 50 (2006): 167-172.

23 K. Shimoi, H. Okada, M. Furugori, T. Goda, S. Takase, M. Suzuki, Y. Hara, H. Yamamoto, and N. Kinae, "Intestinal absorption of luteolin and luteolin 7-O-beta-glucoside in rats and humans," *FEBS Letters* 438 (1998): 220-224.

24 P. Zhou, L. Li, S. Luo, H. Jiang, and S. Zeng, Intestinal absorption of luteolin from peanut hull extract is more efficient than that from individual pure luteolin," *Journal of Agricultural and Food Chemistry* 56 (2008): 296-300.

25 T. Chen, L. Li, X. Lu, H. Jiang, and S. Zeng, "Absorption and excretion of luteolin and apigenin in rats after oral administration of Chrysanthemum morifolium extract," *Journal of Agricultural and Food Chemistry* 55 (2007): 273-277.

26 D. Li, Q. Wang, Z. Yuan, L. Zhang, L. Xu, Y. Cui, and K. Duan, "Pharmacokinetics and tissue distribution study of orientin in rat by liquid chromatography," *Journal of Pharmaceutical and Biomedical Analysis* 47 (2008): 429-434.

27 H. Tamaki, H. Satoh, S. Hori, H. Ohtani, and Y. Sawada, "Inhibitory effects of herbal extracts on breast cancer resistance protein (BCRP) and structure-activity potency relationship of isoflavonoids," *Drug Metabolism and Pharmacokinetics* 25 (2010): 170-179.

28 A. Kawase, Y. Matsumoto, M. Hadano, Y. Ishii, and M. Iwaki, "Differential effects of chrysin on nitrofurantoin pharmacokinetics mediated by intestinal breast cancer resistance protein in rats and mice," *Journal of Pharmacy and Pharmaceutical Sciences* 12 (2009): 150-163.

29 W. Chatuphonpraesert, S. Kondo, K. Jarukamjorn, Y. Kawasaki, T. Sakuma, and N. Nemoto, "Potent modification of inducible CYP1A1 expression by flavonoids," *Biological and Pharmaceutical Bulletin* 33 (2010): 1698-1703.

30 D. Si, Y. Wang, Y. Zhou, Y. Guo, J. Wang, H. Zhou, Z. Li, and J. Fawcett, "Mechanism of CYP2C9 inhibition by flavones and flavonols," *Drug Metabolism and Disposition* 37 (2009): 629-634.

31 T. Sergent, I. Dupont, E. Van der Heiden, M. Scippo, L. Pussemier, Y. Larondelle, and Y. Schneider, "CYP1A1 and CYP3A4 modulation by dietary flavonoids in human intestinal Caco-2 cells," *Toxicology Letters* 191 (2009): 216-222.

32 E. Eaton, U. Walle, A. Lewis, T. Hudson, A. Wilson, and T. Walle, "Flavonoids, potent inhibitors of the human P-form phenolsulfotransferease. Potential role in drug metabolism and chemoprevention," *Drug Metabolism and Disposition* 24 (1996): 232-237.

33 L. Quintieri, P. Palatini, A. Nassi, P. Ruzza, and M. Floreani, "Flavonoids diosmetin and luteolin inhibit midazolam metabolism by human liver microsomes and recombinant CYP3A4 and CYP3A5 enzymes," *Biochemical Pharmacology* 75 (2008): 1426-1437.

34 O. Grundmann, J. Wang, G. P. McGregor, and V. Butterweck, "Anxiolytic activity of a phytochemically characterized Passiflora incarnata extract is mediated via the GABAergic system," *Planta Medica* 74 (2008): 1769-1773.

35 P. de Castro, A. Hoshino, J. da Silva, and F. Mendes, "Possible anxiolytic effect of two extracts of Passiflora quadrangularis L. in experimental models," *Phytotherapy Research* 21 (2007): 481-444.

[36] L. Sena, S. Zucolotto, F. Reginatto, E. Schenkel, and T. Monteiro De Lima, "Neuropharmacological activity of the pericarp of Passiflora edulis flavicarpa Degener: Putative involvement of C-glycosylfavonoids," *Experimental Biology and Medicine* 234 (2009): 967-975.

[37] P. Barbosa, S. Valvassori, C. Bordignon, V. Kappel, M. Martins, E. Gavioli, J. Quevedo, and F. Reginatto, "The aqueous extracts of Passiflora alata and Passiflora edulis reduce anxiety-related behaviors without affecting memory process in rats," *Journal of Medicinal Food* 11 (2008): 282-288.

[38] M. Coleta, M. Batista, M. Campos, R. Carvalho, M. Cotrim, T. Lima, and A. Cunha, "Neuropharmacological evaluation of the putative anxiolytic effects of Passiflora edulis Sims, its subfractions and flavonoid constituents," *Phytotherapy Research* 20 (2006): 1067-1073.

[39] C. Wolfman, H. Viola, A. Paladini, F. Dajas, and J. Medina, "Possible anxiolytic effects of chrysin, a central benzodiazepine receptor ligand isolated from Passiflora coerulea," *Pharmacology, Biochemistry and Behavior* 47 (1994): 1-4.

[40] E. Brown, N. Hurd, S. McCall, and T. Ceremuga, "Evaluation of the anxiolytic effects of chrysin, a Passiflora incarnata extract, in the laboratory rat," *AANA Journal* 75 (2007): 333-337.

[41] P. Zanoli, R. Avallone, and M. Baraldi, "Behavioral characterization of the flavonoids apigenin and chrysin," *Fitoterapia* 71 (2000): S117-123.

[42] J. Deng, Y. Zhou, M. Bai, H. Li, and L. Li, "Anxiolytic and sedative activities of Passiflora edulis f. flavicarpa," *Journal of Ethnopharmacology* 128 (2010): 148-153.

[43] A. Ngan and R. Conduit, "A double-blind, placebo-controlled investigation of the effects of Passiflora incarnata (passionflower) herbal tea on subjective sleep quality," *Phytotherapy Research 2011* 8 (2011): 1153-1159.

[44] L. Krenn, "Passion flower (Passiflora incarnata L.)—a reliable herbal sedative," *WienerMedizinische Wochenschrift* 152 (2002): 404-406.

[45] E. Lakhan and K. F. Vieira, "Nutritional and herbal supplements for anxiety and anxiety-related disorders: systematic review," *Nutrition Journal* 9 (2010): 42-58.

[46] T. T. Faustino, R. B. de Almeida, and R. Andreatini, "Medicinal plants for the treatment of generalized anxiety disorder: a review of controlled clinical studies," *Revista Brasiliera de Psiquiatria* 32 (2010): 429-436.

[47] S. Akhandzadeh, H. Naghavi, M. Vazirian, A. Shayeganpour, H. Rashidi, and M. Khani, "Passionflower in the treatment of generalized anxiety: a pilot double-blind randomized controlled trial with oxazepam," *Journal of Clinical Pharmacy and Therapeutics* 26 (2001): 363-367.

[48] A. Movafegh, R. Alizadeh, F. Hajimohamadi, F. Esfehani, and M. Nejatfar, "Preoperative and oral Passiflora incarnata reduces anxiety in ambulatory surgery patients: a double-blind, placebo-controlled study," *Anesthesia and Analgesia* 106 (2008): 1728-1732.

[49] G. Cravotto, L. Boffa, L. Genzini, and D. Garella, "Phytotherapeutics: an evaluation of the potential of 1000 plants," *Journal of Clinical Pharmacy and Therapeutics* 35 (2010): 11-48.

[50] J. Sarris, A. Panossian, I. Schweitzer, C. Stough, and A. Scholey, "Herbal medicine for depression, anxiety and insomnia: a review of psychopharmacology and clinical evidence," *European Neuorpsychopharmacology* 21 (2011): 841-860.

[51] S. M. Elsas, D. J. Rossi, J. Raber, G. White, C. A. Seeley, W. L. Gregory, C. Mohr, T. Pfankuch, and A. Soumyanath, "Passiflora incarnata L. (Passionflower) extracts elicit GABA currents in hippocampal neurons in vitro, and show anxiogenic and anticonvulsant effects in vivo, varying with extraction method," *Phytomedicine* 17 (2010): 940-949.

[52] B. L. Fiebich, R. Knorle, K. Appel, T. Kammler, and G. Weiss, "Pharmacological studies in an herbal drug combination of St. John's Wort (Hypericum perforatum) and passion flower (Passiflora incarnata): In vitro and in vivo evidence of synergy between Hypericum and Passiflora in antidepressant pharmacological models," *Fitoterapia* 82 (2011): 474-480.

[53] B. Singh, D. Singh, and R. Goel, "Dual protective effect of Passiflora incarnata in epilepsy and associated post-ictal depression," *Journal of Ethnopharmacology* 139 (2012): 273-279.

[54] M. Nassiri-Asi, S. Shariati-Rad, and F. Zamansoltani, "Anticonvulsant effects of aerial parts of Passiflora incarnata extract in mice: involvement of benzodiazepine and opioid receptors," *BMC Complementary and Alternative Medicine* 7 (2007): 26.

[55] R. Da Silva, R. Yunes, M. de Souza, F. Delle Monache, and V. Cechinel-Filho, "Antinociceptive properties of conocarpan and orientin obtained from Piper solmsianum C. DC. var. solmsianum (Piperaceae)," *Journal of Natural Medicines* 64 (2010): 402-408.

[56] M. Rylski, H. Duriasz-Rowinska, and W. Rewerski, "The analgesic action of some flavonoids in the hot plate test," *Acta Physiologica Polonica* 30 (1979): 385-388.

[57] A. C. Paladini, M. Marder, H. Viola, C. Wolfman, C. Wasowski, and J. H. Medina, "Flavonoids and the central nervous system: from forgotten factors to potent anxiolytic compounds," *Journal of Pharmacy and Pharmacology* 51 (1999): 519-526.

[58] J. Medina, A. Paladini, C. Wolfman, M. Levi de Stein, D. Calvo, L. Diaz, and C. Pena, "Chrysin (5,7-di-OH-flavone), a naturally-occurring ligand for benzodiazepine receptors, with anticonvulsant properties," *Biochemical Pharmacology* 40 (1990): 2227-2231.

[59] T. Ichimura, A. Yamanaka, T. Ichiba, T. Toyokawa, Y. Kamada, T. Tamamura, and S. Maruyama, "Antihypertensive effect of an extract of Passiflora edulis rind in spontaneously hypertensive rats," *Bioscience, Biotechnology and Biochemistry* 70 (2006): 18-21.

[60] J. Kellis and L. Vickery, "Inhibition of human estrogen synthetase (aromatase) by flavones," *Science* 225 (1984): 1032-1034.

61 N. Ta and T. Walle, "Aromatase inhibition by bioavailable methylated flavones," *Journal of Steroid Biochemistry and Molecular Biology* 107 (2007): 127-129.

62 T. Walle, Y. Otake, J. Brubaker, U. Walle, and P. Halushka, "Disposition and metabolism of the flavonoid chrysin in normal volunteers," *British Journal of Clinical Pharmacology* 51 (2001): 143-146.

63 C. Gambelunghe, R. Rossi, M. Sommavilla, C. Ferranti, R. Rossi, C. Ciculi, S. Gizzi, A. Micheletti, and S. Rufini, "Effects of chrysin on urinary testosterone in human males," *Journal of Medicinal Food* 6 (2003): 387-390.

64 N. Saarinen, S. Joshi, M. Ahotupa, X. Li, J. Ammala, S. Makela, and R. Santti, "No evidence for the in vivo activity of aromatase-inhibiting flavonoids," *Journal of Steroid Biochemistry and Molecular Biology* 78 (2001): 231-239.

65 C. M. Lin, S. T. Huang, Y. C. Liang, M. S. Lin, C. M. Shih, Y. C. Chang, T. Y. Chen, and C. T. Chen, "Isovitexin suppresses lipopolysaccharide-mediated inducible nitric oxide synthase through inhibition of NF-kappa B in mouse macrophages," *Planta Medica* 71 (2005): 748-753.

66 A. Montanher, S. Zucolotto, E. Schenkel, and T. Frode, "Evidence of anti-inflammatory effects of Passiflora edulis in an inflammation model," *Journal of Ethnopharmacology* 109 (2007): 281-288.

67 L. Dong, Y. Fan, X. Shao, and Z. Chen, "Vitexin protects against myocardial ischemia/reperfusion injury in Langendorff-perfused rat hearts by attenuating inflammatory response and apotosis," *Food Chemistry and Toxicology* 49 (2011): 3211-3216.

68 R. Sathish, A. Sahu, and K. Natarajan, "Antiulcer and antioxidant activity of ethanolic extract of Passiflora foetida L.," *Indian Journal of Pharmacology* 43 (2011): 336-339.

69 E. Shin, H. Kwon, Y. Kim, H. Shin, and J. Kim, "Chrysin, a natural flavone, improves murine inflammatory bowel diseases," *Biochemical, Biophysical Research Communications* 381 (2009): 502-507.

70 Y. Bae, S. Lee, and S. Kim, "Chrysin suppresses mast cell-mediated allergic inflammation: involvement of calcium, caspase-1 and nuclear factor-κB," *Toxicology and Applied Pharmacology* 254 (2011): 56-64.

71 M. C. Pascoe, S. G. Crewther, L. M. Carey, and D. P. Crewther, "What you eat is what you are—A role for polyunsaturated fatty acids in neuroinflammation induced depression?" *Clinical Nutrition* 30 (2011): 407-415.

72 E. Pichichero, R. Cicconi, M. Mattei, and A. Canini, "Chrysin-induced apoptosis is mediated through p38 and Bax activation in B16-F1 and A375 melanoma cells," *International Journal of Oncology* 38 (2011): 473-483.

73 H. Choi, J. Eun, B. Kim, S. Kim, H. Jeon, and Y. Soh, "Vitexin, an HIF-1alpha inhibitor, has anti-metastatic potential in PC12 cells," *Molecular Cells* 22 (2006): 291-299.

[74] M. C. Carrasco, J. R. Vallejo, M. Pardo-de-Santayana, D. Peral, M. A. Martin, and J. Altimiras, "Interactions of Valeriana Officinalis L. and Passiflora incarnata L. in a patient treated with lorazepam," *Phytotherapy Research* 23 (2009): 1795-1796.

## Chapter 14

[1] S. C. Li, *Chinese Medicinal Herbs, A Modern Edition of a Classic Sixteenth Century Manual,* trans., F. P. Smith and G. A. Stuart (Mineola, NY: Dover Publications, Inc., 2003).

[2] M. S. Premila, *Ayurvedic Herbs, A Clinical Guide to the Healing Plants of Traditional Indian Medicine* (New York: The Haworth Press, Inc., 2006).

[3] O. Usmani, M. Belvisi, H. Patel, N. Crispino, M. Birrell, M. Korbonits, D. Korbonits, and P. Barnes, "Theobromine inhibits sensory nerve activation and cough," *FASEB Journal* 19 (2004): 231-233.

[4] M. Lucas, F. Mirzaei, A. Pan, O. I. Okereke, W. Willett, E. O' Reilly, K. Koenen, and A. Ascherio, "Coffee, caffeine, and risk of depression among women," *Archives of Internal Medicine* 171 (2011): 1571-1578.

[5] V. Cropley, R. Croft, B. Silber, C. Neale, A. Scholey, C. Stough, and J. Schmitt, "Does coffee enriched with chlorogenic acids improve mood and cognition after acute administration in healthy elderly? A pilot study," *Psychopharmacology (Berlin)* 21 (2012): 737-749.

[6] D. Ganmaa, W. Willett, Y. Li, D. Feskanich, R. van Dam, E. Lopez-Garcia, D. Hunter, and M. Holmes, "Coffee, tea, caffeine and risk of breast cancer: a 22-year follow-up," *International Journal of Cancer* 122 (2008): 2071-2076.

[7] A. Tavani and C. La Vecchia, "Coffee: decaffeinated coffee, tea and cancer of the colon and rectum: a review of epidemiological studies," *Cancer Causes and Control* 15 (2004): 743-757.

[8] R. M. van Dam and F. B. Hu, "Coffee consumption and risk of type 2 diabetes: a systematic review," *Journal of the American Medical Association* 294 (2005): 97-104.

[9] R. M. van Dam, "Coffee and type 2 diabetes: from bean to beta-cells," *Nutrition Metabolism and Cardiovascular Disease* 16 (2006): 69-77.

[10] E. Lopez-Garcia, R. M. van Dam, T. Y. Li, F. Rodriquez-Artalejo, and F. B. Hu, "The relationship of coffee consumption with mortality," *Annals of Internal Medicine* 148 (2008): 904-914.

[11] K. A. Grove and J. D. Lambert, "Laboratory, epidemiological and human intervention studies show that tea (Camellia sinensis) may be useful in the prevention of obesity," *Journal of Nutrition* 140 (2010): 446-453.

[12] Y. S. Lin, Y. J. Tsai, J. S. Tsay, and J. K. Lin, "Factors affecting the levels of tea polyphenols and caffeine in tea leaves," *Journal of Agricultural Food Chemicals* 51 (2003): 1864-1873.

[13] U. J. Unachukwu, S. Ahmed, A. Kavalier, J. T. Lyles, and E. J. Kennelly, "White and green teas (Camillia sinensis var. sinensis): variation in phenolic, methylxanthine and antioxidant profiles," *Journal of Food Science* 75(2010): C541-C548.

[14] D. do Carmo Carvalho, M. R. Brigagao, M. H. dos Santos, F. B. de Paula, A. Giusti-Paiwa, and L. Azevedo, "Organic and conventional Coffea arabica L.: a comparative study of the chemical composition and physiological, biochemical and toxicological effects in Wistar rats," *Plant Foods for Human Nutrition* 66 (2011): 114-121.

[15] M. Rusconi and A. Conti, "Theobroma cacao L., the food of the gods: a scientific approach beyond myths and claims," *Pharmacological Research* 61 (2010): 5-13.

[16] T. Stark and T. Hofmann, "Isolation, structure determination, synthesis and sensory activity of N-phenylpropenoyl-L-amino acids from cocoa (Theobroma cacao)," *Journal of Agricultural Food Chemistry* 53 (2005): 5419-5428.

[17] M. Natsume, N. Osakabe, M. Yamagishi, T. Takizawa, T. Nakamura, H. Miyatake, T. Hatano, and T. Yoshida, "Analyses of polyphenols in cacao liquor, cocoa and chocolate by normal-phase and reversed-phase HPLC," *Bioscience, Biotechnology and Biochemistry* 64 (2000): 2581-2587.

[18] K. A. Berte, M. R. Beux, P. K. Spada, M. Salvador, and R. Hoffmann-Ribani, "Chemical composition and antioxidant activity of yerba-mate (Ilex paraguariensis A. St.-Hil., Aquifoliaceae) extract as obtained by spray drying," *Journal of Agricultural Food Chemistry* 59 (2011): 5523-5527.

[19] C. I. Heck and E. G. de Mejia, "Yerbe mate tea (Ilex paraguariensis): a comprehensive review on chemistry, health implications and technological considerations," *Journal of Food Science* 72 (2007): R138-R151.

[20] Y. Wang and C. E. Lau, "Caffeine has similar pharmacokinetics and behavioral effects via the i.p. and p.o. routes of administration," *Pharmacology, Biochemistry and Behavior* 60 (1998): 271-278.

[21] A. Lelo, D. Birkett, R. Robson, and J. Miners, "Comparative pharmacokinetics of caffeine and its primary demethylated metabolites paraxanthine, theobromine and theophylline in man," *British Journal of Clinical Pharmacology* 22 (1986): 177-182.

[22] T. W. Rall, "Central Nervous System Stimulants: the Xanthines," in *The Pharmacological Basis of Therapeutics,* 6th edition, eds., A. G. Gilman, L. Goodman, and A. Gilman (New York: Macmillan Publishing Co., Inc., 1980).

[23] H. Cornish and A. Christman, "A study of the metabolism of theobromine, theophylline, and caffeine in man," *Journal of Biological Chemistry* 228 (1957): 315-323.

[24] D. C. Zhang, "Pharmacological effect of caffeine and related purine alkaloids," in *Tea, Bioactivity and Therapeutic Potential*, ed., Y-S. Zhen (New York: Taylor and Francis, 2002).

25 T. Bakuradze, N. Boehm, C. Janzowski, R. Lang, T. Hofmann, J. P. Stockis, F. W. Albert, H. Stiebitz, G. Bytof, I. Lantz, M. Baum, and G. Eisenbrand, "Antioxidant-rich coffee reduces DNA damage, elevates glutathione status and contributes to weight control: results from an intervention study," *Molecular Nutrition and Food Research* 55 (2011): 793-797.

26 T. Taguri, T. Tanaka, and I. Kouno, "Antibacterial spectrum of plant polyphenols and extracts depending upon hydroxyphenyl structure," *Biological & Pharmaceutical Bulletin* 29 (2006): 2226-2235.

27 R. Tsao, "Chemistry and biochemistry of dietary polyphenols, *Nutrients* 2 (2010): 1231-1246.

28 S. Ferre, "An update on the mechanisms of the psychostimulant effects of caffeine," *Journal of Neurochemistry* 105 (2008): 1067-1079.

29 S. J. Grant and D. E. Redmond, Jr., "Methylxanthine activation of noradrenergic unit activity and reversal by clonidine," *European Journal of Pharmacology* 85 (1982): 105-109.

30 G. R. Stoner, L. R. Skirboll, S. Werkman, and D. W. Hommer, "Preferential effects of caffeine on limbic and cortical dopamine systems," *Biological Psychiatry* 23 (1988): 761-768.

31 D. G. Rainnie, H. C. Grunze, R. W. McCarley, and R. W. Greene, "Adenosine inhibition of mesopontine cholinergic neurons: implications for EEG arousal," *Science* 263: (1994): 689-692.

32 S. Ferre, G. Von Euler, B. Johansson, and B. B. Fredholm, "Adenosine-dopamine interactions in the brain," *Neuroscience* 511 (1992): 501-512.

33 M. Lazarus, H. Y. Shen, Y. Cherasse, W. M. Qu, Z. L. Huang, C. E. Bass, R. Winsky-Sommerer, K. Semba, B. B. Fredholm, D. Boison, O. Hayaishi, Y. Urade, and J. F. Chen, "Arousal effect of caffeine depends on adenosine A2A receptors in the shell of the nucleus accumbens," *Journal of Neuroscience* 31 (2011): 10067-10075.

34 B. B. Fredholm, K. Battig, J. Holmen, A. Nehlig, and E. E. Zvartau, "Actions of caffeine in the brain with special reference to factors that contribute to its widespread use," *Pharmacological Reviews* 51 (1999): 83-133.

35 S. A. Mandel, Y. Avramovich-Tirosh, L. Reznichenko, H. Zheng, O. Weinreb, T. Amit, and M. B. Youdim, "Multifunctional activities of green tea catechins in neuroprotection. Modulation of cell survival genes, iron-dependent oxidative stress and PKC signaling pathway," *Neurosignals* 14 (2005): 46-60.

36 E.L. Schiffrin, "Antioxidants in hypertension and cardiovascular disease," *Molecular Interventions* 10 (2010): 354-362.

37 M. Kimura, K. Umegaki, Y. Kasuya, A. Sugisawa, and M. Higuchi, "The relation between single/double or repeated tea catechin ingestions and plasma antioxidant activities in humans," *European Journal of Clinical Nutrition* 56 (2002): 1186-1193.

[38] A.E. Koutelidakis, K. Argiri, M. Serafini, C. Proestos, M. Komaitis, M. Pecorari and M. Kapsokefalou, "Green tea, white tea and Pelargonium purpureum increase the antioxidant capacity of plasma and some organs in mice," *Nutrition* 25 (2009): 453-458.

[39] N. Khan and H. Mukhtar, "Tea polyphenols for health promotion," *Life Sciences* 81 (2007): 519-533.

[40] K. Boehm, F. Borrelli, E. Ernst, G. Habacher, S. K. Hung, S. Milazzo, and M. Horneber, "Green tea (Camellia sinensis) for the prevention of cancer," *Cochran Database of Systematic Reviews* 2009, Issue 3. Art. No.: CD005004. DOI: 10.1002/14651858.CD005004.pub2.

[41] L. Elbling, R. M. Weiss, O. Teufelhofer, M. Uhl, S. Knasmueller, R. Schulte-Hermann, W. Berger, and M. Micksche, "Green tea extract and (-)-epigallocatechin -3-gallate, the major tea catechin, exert oxidant but lack antioxidant activities," *FASEB Journal* 19 (2005): 807-809.

[42] J. D. Lambert and R. J. Elias, "The antioxidant and pro-oxidant activities of green tea polyphenols: a role in cancer prevention," *Archives of Biochemistry and Biophysics* 501 (2010): 65-72.

## Chapter 15

[1] H. B. Murphree, "Opium Alkaloids," in *Drill's Pharmacology in Medicine*, J. R. DiPalma, ed., 3rd edition (New York: McGraw-Hill Book Co., 1965).

[2] S. B. Karch, "Ma Huang and the Ephedra Alkaloids," in *Herbal Products*, T. S. Tracy and R. L. Kingston, eds., 2nd edition (Towota, NJ: Humana Press, Inc., 2007).

[3] N. Weiner, "Norepinephrine, Epinephrine, and the Sympathomimetic Amines," in *Goodman and Gilman's the Pharmacological Basis of Therapeutics*, A. G. Gilman, L. S. Goodman, and A. Gilman, eds., 6th edition (New York: Macmillan Publishing Co., 1980).

[4] E. P. Claus, *Gathercoal and Wirth, Pharmacognosy*, 3rd edition (Philadelphia: Lea and Febiger, 1956).

[5] R. C. Clarke and D. P. Watson, "Cannabis and Natural Cannabis Medicines," in *Marijuana and the Cannabinoids*, M. A. El Sohly, ed., (Totowa, NJ: Humana Press, Inc. 2007).

[6] R. Pertwee, "Receptors and pharmacodynamics: natural and synthetic cannabinoids and endocannabinoids," in *The Medicinal Uses of Cannabis and Cannabinoids*, G. W. Guy, B. A. Whittle, and P. J. Robson, eds., (London: Pharmaceutical Press, 2004).

# Index

## A

## B

## C

## D

## E



## F

## G

## H

## I



## J-K



## L

## M

## N

## O



## P

## Q-R



## S

## T

## U



## V



## W-X-Y-Z

The life sciences revolution is transforming
our world as profoundly as the industrial
and information revolutions did
in the last two centuries.
FT Press Science will capture the excitement and
promise of the new life sciences, bringing breakthrough
knowledge to every professional and interested citizen.
We will publish tomorrow's indispensable work in
genetics, evolution, neuroscience, medicine,
biotech, environmental science, and whatever
new fields emerge next.
We hope to help you make sense of the future,
so you can *live* it, *profit* from it, and *lead* it.